Lecture Notes in Mathematics

Edited by A. Dold and B. Eckmann

Subseries: Tata Institute of Fundamental Research, Bombay
Adviser: M. S. Narasimhan

302

Michel Demazure

Lectures on p-Divisible Groups

Springer-Verlag
Berlin Heidelberg New York Tokyo

Author

Michel Demazure
Centre de Mathématique, Ecole Polytechnique
91128 Palaiseau Cedex, France

1st Edition 1972
2nd Printing 1986

Mathematics Subject Classification (1970): 14-02, 14L05

ISBN 3-540-06092-8 Springer-Verlag Berlin Heidelberg New York Tokyo
ISBN 0-387-06092-8 Springer-Verlag New York Heidelberg Berlin Tokyo

Printing and binding: Beltz Offsetdruck, Hemsbach/Bergstr.
2146/3140-543210

Lectures on p-divisible groups

The aim of these lectures, given at the Tata Institute in January - February 1971, was to explain the contents of chapters 1, 2 and 4 of

MANIN (I), The theory of commutative formal groups over fields of finite characteristic, English Trans., Russian Math. Sur. 18.

For general facts about algebraic groups, our reference is

DEMAZURE (M) and GABRIEL (P), Groupes algébriques, Tome 1, North Holland Pub. Co., 1970, which shall be abbreviated as D.G.

For supplementary material the reader may refer to:

HONDA (T), Isogeny classes of abelian varieties over finite fields, J. Math. Soc. Jap., 20, 83-95, (1968).

HONDA (T), On the theory of commutative formal groups, J. Math. Soc. Jap., 22 213-246, (1970).

TATE (J), p-divisible groups ; Proceedings of a conference on local fields, (Driebergen 1966) Springer-Verlag, 1967.

TATE (J), Classes d'isogénies de variétés abéliennes sur un corps fini(d'après T. HONDA), Seminaire Bourbaki, 352, Nov. 1968, Paris multigraphé.

TATE (J), Endomorphisms of abelian varieties over finite fields, Inventiones Math., 2 134-144 (1966)

N.B. The typing of these notes was done by Mr.P.Joseph, of the Tata Institute. He did a very good job.

<u>Notational Conventions</u>. If $\underline{C}$ is a category, and A an object of $\underline{C}$, we write simply $A \in \underline{C}$; if $A, B \in \underline{C}$, the set of morphisms in $\underline{C}$ of A to B is denoted by $\underline{C}(A,B)$.

By a ring we always mean, if not otherwise stated, a commutative ring with unit.

M. Demazure

Table of Contents

CHAPTER I

SCHEMES AND FORMAL SCHEMES

1. k-functors

Let k be a ring and $\underline{M}_k$ be the category of k-rings (i.e. commutative associative k-algebras with unit, or simply couples (R,φ) where R is a ring and $\varphi: k \longrightarrow R$ a morphism). Actually, for set-theoretical reasons, one should not take the category of all k-rings, but a smaller one (see D.G. page XXV-XXVI) but we shall not bother about this point.

A k-functor is by definition a covariant functor from $\underline{M}_k$ to the category $\underline{E}$ of sets; the category of k-functors is denoted by $\underline{M}_k\underline{E}$.

Example. The affine line $\underline{O}_k$ is defined by $\underline{O}_k(R) = R$, $R \in \underline{M}_k$.

If $\varphi: R \longrightarrow S$ is an arrow of $\underline{M}_k$, if $X \in \underline{M}_k\underline{E}$, and if $x \in X(R)$, we shall write x_S (or sometimes x) instead of $X(\varphi)(x) \in X(S)$; if $f: X \longrightarrow Y$ is an arrow of $\underline{M}_k\underline{E}$, if $R \in \underline{M}_k$ and $x \in X(R)$, we shall write $f(x)$ instead of $f(R)(x) \in Y(R)$; with these notations, the fact that f is a morphism of functors amounts to $f(x)_S = f(x_S)$.

The category $\underline{M}_k\underline{E}$ has projective limits, for example:

a) a final object e is defined by $e(R) = \{\emptyset\}$, $R \in \underline{M}_k$,

b) if $X, Y \in \underline{M}_k\underline{E}$, the product $X \times Y$ is defined by $(X \times Y)(R) = X(R) \times Y(R)$,

c) if $X \xrightarrow{f} Z \xleftarrow{g} Y$ is a diagram of $\underline{M}_k\underline{E}$, the <u>fibre product</u> $T = X \times_Z Y$ is defined by $T(R) = X(R) \times_{Z(R)} Y(R) = \{(x,y) \in X(R) \times Y(R),\ f(x) = g(y)\}$; more generally, one has $(\varprojlim X_i)(R) = \varprojlim X_i(R)$,

d) $f: X \longrightarrow Y$ is a <u>monomorphism</u> if and only if $f(R): X(R) \longrightarrow Y(R)$ is injective for each R. We say that X is a subfunctor of Y if $X(R) \subset Y(R)$ and $f(R)$ is the inclusion, for all R.

Let $k' \in \underline{M}_k$; as any k'-algebra can be viewed as a k-algebra, there is an obvious functor $\underline{M}_{k'} \longrightarrow \underline{M}_k$ and therefore an obvious functor $\underline{M}_k\underline{E} \longrightarrow \underline{M}_{k'}\underline{E}$; the latter is denoted by $X \longrightarrow X \otimes_k k'$. So, if R is a k'-ring and $R_{[k]}$ the underlying k-ring, one has

$$X \otimes_k k'(R) = X(R_{[k]});$$

the functor $X \longrightarrow X \otimes_k k'$ is called the <u>base-change</u> functor or <u>scalar-extension</u> functor. It commutes with projective limits, hence is <u>left-exact</u>.

For instance $\underline{O}_k \otimes_k k'$ can be (and will be) identified with $\underline{O}_{k'}$.

2. <u>Affine</u> k-<u>schemes</u>.

Let $A \in \underline{M}_k$; the k-functor $Sp_k A$ (or simply $Sp\ A$) is defined by

$$Sp_k A(R) = \underline{M}_k(A,R)$$

$$Sp_k A(\varphi) = \{\psi \longmapsto \varphi \circ \psi\} \text{ for } \varphi: R \longrightarrow S;$$

if $f: A \longrightarrow B$ is an arrow of $\underline{M}_k$, then $Sp_k f: Sp_k B \longrightarrow Sp_k A$ is obviously defined. So $A \longmapsto Sp_k A$ is a contravariant functor from $\underline{M}_k$ to $\underline{M}_k\underline{E}$.

An <u>affine</u> k-<u>scheme</u> is a k-functor isomorphic to a $Sp_k A$. For instance $\underline{O}_k$ is an affine k-scheme because

$$Sp_k k[T](R) = \underline{M}_k(k[T],R) \simeq R = \underline{O}_k(R).$$

Let X be a k-functor and A a k-ring. We have the very simple and very important Yoneda bijection

$$\underline{M}_k\underline{E}(Sp_k A, X) \xrightarrow{\sim} X(A) \ ;$$

to $f: Sp_k A \longrightarrow X$ is associated $\xi = f(id_A) \in X(A)$; conversely, if $\xi \in X(A)$ and $\varphi \in Sp_k(A)(R) = \underline{M}_k(A,R)$, we put $f(\varphi) = X(\varphi)(\xi)$; with our notation, the correspondence between f and ξ is simply $f(\varphi) = \varphi(\xi)$.

As an example, we take $X = Sp_k B$; then $X(A) = \underline{M}_k(B,A)$, and we have a bijection

$$\underline{M}_k\underline{E}(Sp_k A, Sp_k B) \simeq M_k(B,A) \ ;$$

it means that $A \longrightarrow Sp_k A$ is fully faithful, or equivalently that it induces an anti-equivalence between the category of k-rings and the category of affine k-schemes.

This fundamental equivalence can also be looked at in the following way: Let X be any k-functor; define a function on X to be a morphism $f: X \longrightarrow \underline{O}_k$, i.e. a functorial system of maps $X(R) \longrightarrow R$. The set of these functions, say $O(X)$, has an obvious k-ring structure: if f, $g \in O(X)$, $\lambda \in k$, then

$$(f+g)(x) = f(x) + g(x)$$
$$(fg)(x) = f(x)g(x)$$
$$(\lambda f)(x) = \lambda f(x)$$

for any $R \in \underline{M}_k$ and any $x \in X(R)$. If $x \in X(R)$ is fixed, then by the very definition of the k-ring structure of $O(X)$, $f \mapsto f(x)$ is an element of $\underline{M}_k(O(X),R) = Sp\ O(X)$; we therefore have a canonical morphism

$$\alpha: X \longrightarrow Sp\ O(X) \ .$$

It is easily seen that α is universal with respect to morphisms of X into affine k-schemes (any such morphism can be uniquely factorized through α). The definition of affine k-schemes can be rephrased as: X is an affine k-scheme if and only if

α is an isomorphism. For instance $O(\underline{O}_k)$ is the polynomial algebra $k[T]$ generated by the identity morphism $T: \underline{O}_k \longrightarrow \underline{O}_k$.

The functor $A \longrightarrow Sp_k A$ commutes with projective limits and base-change: one has the following obvious isomorphisms:

$$Sp(A) \times_{Sp(C)} Sp(B) \simeq Sp(A \otimes_C B)$$

$$\varprojlim Sp(A_i) \simeq Sp(\varinjlim A_i)$$

$$Sp_k(A) \otimes_k k' \simeq Sp_k(A \otimes_k k'),$$

(the last one explaining the notation $\otimes$ for base-change); as a consequence, the full subcategory of affine schemes is stable under projective limits and base-change.

3. Closed and open subfunctors; schemes.

Let X be a k-functor and E be a set of functions on $X; E \subset O(X)$. We define two subfunctors $V(E)$ and $D(E)$ of X:

$$V(E)(R) = \{x \in X(R) \mid f(x) = 0 \text{ for all } f \in E\}.$$

$$D(E)(R) = \{x \in X(R) \mid f(x) \text{ for } f \in E, \text{ generate the unit ideal of } R\}.$$

If $u: Y \to X$ is a morphism of k-functors and $F = \{f \circ u, f \in E\} \subset O(Y)$, then $u^{-1}(V(E)) = V(F)$, $u^{-1}(D(E)) = D(F)$ [if $u: Y \longrightarrow X$ is a morphism of k-functors and Z is a subfunctor of X, then $u^{-1}(Z)$ is defined as the subfunctor of Y such that $u^{-1}(Z)(R) = \{y \in Y(R) \mid u(y) \in Z(R)\}$].

If X is an affine k-scheme, then

1) $V(E)$ is an affine k-scheme with $O(V(E)) = O(X)/E\, O(X)$

2) if $E = \{f\}$ has only one element, then $D(E)$ is an affine k-scheme with $O(D(\{f\}) = O(X)[f^{-1}] = O(X)[T]/(Tf-1)$.

Proof. If $X = Sp\ A$, and $E \subset A = O(X)$, then for all $R \in \underline{M}_k$,

$$V(E)(R) = \{\varphi \in \underline{M}_k(A,R) \mid \varphi(E) = 0\} \simeq \underline{M}_k(A/EA, R)$$

$$D(\{f\})(R) = \{\varphi \in \underline{M}_k(A,R) \mid \varphi(f) \text{ is invertible}\} \simeq \underline{M}_k(A[f^{-1}], R) .$$

Definition. The subfunctor Y of X is said to be closed (resp. open) if for any morphism $u: T \longrightarrow X$ where T is an affine scheme, the subfunctor $u^{-1}(Y)$ of T is of the form $V(E)$ (resp. $D(E)$).

For instance, if X is affine, then Y is closed (resp. open) if and only if it is a $V(E)$ (resp. $D(E)$). As a corollary, a closed subfunctor of an affine k-scheme is also an affine k-scheme; this need not be true for open subfunctors: take $X = \mathrm{Sp}\, k[T,T'] \simeq \underline{O}_k^2$ and $Y = D(\{T,T'\})$. In the functorial setting, the precise definition of a not-necessarily affine k-scheme is a bit complicated. Let us give it for the sake of completeness:

Definition. The k-functor X is a scheme if:

1) it is a "local" k-functor: for any k-ring R and any "partition of unity" f_i of R(= family of elements of R such that $\sum Rf_i = R$), given elements $x_i \in X(R[f_i^{-1}])$ such that the images of x_i and x_j in $X(R[f_i^{-1} f_j^{-1}])$ coincide for all couples (i,j), then there exists one and only one $x \in X(R)$ which maps on to the x_i.

2) There exists a family (U_j) of open subfunctors with the following properties: each U_j is an affine k-functor; for any field $K \in \underline{M}_k$, $X(K)$ is the union of the $U_j(K)$.

From this definition follows easily:

Proposition 1) an open or closed subfunctor of a k-scheme is a k-scheme,

2) any finite projective limit (e.g. fibre product) of k-schemes is a k-scheme,

3) if X is a k-scheme, then $X \otimes_k k'$ is a k'-scheme.

As an illustration of 1), let $A \in \underline{M}_k$ and $E \subset A$; then $D(E) \subset \mathrm{Sp}\, A$ is a k-scheme, because it is local and covered by the affine k-schemes $D(\{f\})$, $f \in E$. Also note that the limit of a directed projective system of schemes is not generally a scheme (although it is in the affine case, as already seen).

4. The geometric point of view.

Let X be a k-functor; we want to define a geometric space (topological space with a sheaf of local rings) $|X|$ associated to X. First, the underlying set of $|X|$ is defined as follows: a point of $|X|$ is an equivalence class of elements of all $X(K)$ where K runs through the fields of $\underline{M}_k$, $x \in X(K)$ and $x' \in X(K')$ being equivalent if there exist two morphisms of $\underline{M}_k$, say $K \to L$, $K' \to L$, where L is a field, with $x_L = x'_L$.

Second, the topology. If Y is a subfunctor of X, then $|Y|$ can be identified with a subset of $|X|$; we define a subset U of $|X|$ to be open if there exists an open subfunctor Y of X, such that $|Y| = U$; moreover, such a Y can be proved to be unique, and we write $Y = X_U$.

Third, the sheaf is the associated sheaf to the presheaf of rings $U \longrightarrow O(X_U)$.

As an example, take $X = \mathrm{Sp}\, A$, $A \in \underline{M}_k$. Then $|\mathrm{Sp}\, A|$ is the usual spectrum $\mathrm{Spec}\, A$ of A: the points of $\mathrm{Spec}\, A$ are the prime ideals of A; the open sets are the $|D(S)| = \{p \mid S \not\subset p\}$, $S \subset A$; the sheaf is associated to the presheaf $|D(S)| \longrightarrow A[S^{-1}]$. (One basic theorem asserts that the ring of sections of the sheaf over $|D(\{f\})|$ is $A[f^{-1}]$).

In the general case, for all $A \in \underline{M}_k$, and all $\xi \in X(A)$, the Yoneda morphism $\mathrm{Sp}\, A \longrightarrow X$ associated to ξ defines a ringed-space-morphism $\mathrm{Spec}\, A \longrightarrow |X|$ and $|X|$ can be proved to be the inductive limit of the (non-directed) system of the $\mathrm{Spec}\, A$. (D.G. I, § 1, n°4).

One has then the following comparison theorem (D.G.I, §1,4.4.)

Theorem. $X \longrightarrow |X|$ induces an equivalence between the category of k-schemes and the category of geometric spaces locally isomorphic to a Spec A, $A \in \underline{M}_k$.

One can give a quasi-inverse functor: there is a functorial bijection between $X(R)$ and the set of geometric-space-morphisms from Spec R to $|X|$, as follows from the theorem and Yoneda's isomorphism.

By this equivalence, one defines geometric objects associated to the k-scheme X : the local rings $O_{X,x}$ and the residue fields $\kappa(x)$, $x \in |X|$; all are k-rings.

5. Finiteness conditions.

Let k be a field. A k-scheme X is said to be finite if it is affine and if $O(X)$ is a finite dimensional vector space; if X is finite, then $[O(X):k]$ is called the rank $rk(X)$ of X. A k-scheme X is locally algebraic (algebraic) if it has a covering (a finite covering) by open subfunctors X_i which are affine k-schemes such that each $O(X_i)$ is a finitely generated k-algebra. If X is an affine k-scheme, then the following conditions are equivalent:

1) X is algebraic, 2) X is locally algebraic, 3) $O(X)$ is a finitely generated k-algebra (D.G.I, §3,1.7).

It follows from the Normalization lemma that X is finite if and only if X is algebraic and $|X|$ finite. It follows from the Nullstellensatz that if X is locally algebraic and $\neq \emptyset$ (one defines $\emptyset(R) = \emptyset$ for all R, or equivalently $|\emptyset| = \emptyset$), then $X(K) \neq \emptyset$ for some finite extension K of k. Let X be a (locally) algebraic k-scheme, k algebraically closed; then if U is an open subfunctor of X, $U(k) = \emptyset$ implies $U = \emptyset$. This easily implies that if one views $X(k)$ as the subspace of $|X|$ whose points are the $x \in |X|$ such that $\kappa(x) = k$, the open subsets of $|X|$ and the open subsets of $X(k)$ are in a bijective correspondence (by $|U| \longrightarrow U(k)$).

It is therefore equivalent to know the k-scheme X, or the k-geometric space $X(k)$ - the only difference between the $X(k)$'s and Serre's algebraic spaces lies in that the latter have no nilpotent elements in their local rings, whereas the former may have. As we shall see later on, this is an important difference. Serre's algebraic spaces correspond to "reduced" algebraic k-schemes (i.e. with no nilpotent elements). A similar discussion can be made in the case of a general field k; one has to replace $X(k)$ by the set of closed points of $|X|$ (by the Nullstellensatz, $x \in |X|$ is closed if and only if $\kappa(x)$ is a finite extension of k).

6. The four definitions of formal schemes.

From now on, k is assumed to be a field.

We denote by $\underline{Mf}_k$ the full subcategory of $\underline{M}_k$ consisting of finite (= finite dimensional) k-rings. A k-formal functor is a covariant functor $F: \underline{Mf}_k \to \underline{E}$; the category of k-formal functors is denoted by $\underline{Mf}_k\underline{E}$; this category has finite projective limits. The inclusion functor $\underline{Mf}_k \longrightarrow \underline{M}_k$ gives a canonical functor $\underline{M}_k\underline{E} \longrightarrow \underline{Mf}_k\underline{E}$ called the completion functor: if $X \in \underline{M}_k\underline{E}$, then $\hat{X} \in \underline{Mf}_k\underline{E}$ is defined by $\hat{X}(R) = X(R)$ for $R \in \underline{Mf}_k$. The completion-functor is obviously left-exact.

If $A \in \underline{Mf}_k$, we denote by $\mathrm{Spf}_k\, A$ or $\mathrm{Spf}\, A$ the k-formal-functor $R \longrightarrow \underline{Mf}_k(A,R)$; one has obviously $\widehat{\mathrm{Sp}\, A} = \mathrm{Spf}\, A$, and for any $F \in \underline{Mf}_k\underline{E}$ a Yoneda isomorphism $\underline{Mf}_k\underline{E}(\mathrm{Spf}\, A, F) \xrightarrow{\sim} F(A)$, $A \in \underline{Mf}_k$. In particular, the functor $A \longrightarrow \mathrm{Spf}\, A$ is fully-faithful, or, what amounts to the same, the functor $X \to \hat{X}$, X a finite k-scheme, is fully faithful. We therefore can view the category of finite k-schemes as a full subcategory of either $\underline{M}_k\underline{E}$ or $\underline{Mf}_k\underline{E}$ (we shall say: "the completion does not change the finite k-schemes").

a) By definition, a k-formal-scheme is a k-formal functor which is the limit of a directed inductive system of finite k-schemes: F is a k-formal-scheme if there exists a directed projective system (A_i) of finite k-rings and functorial (in R) isomorphisms:

$$F(R) \simeq \lim \underline{Mf}_k(A_i,R) = \varinjlim \mathrm{Spf}\,(A_i)(R)$$

For any k-formal functor G, one has a Yoneda isomorphism

$$\underline{Mf_k E}(\varinjlim \mathrm{Spf}(A_i),\ G) = \varprojlim\ G(A_i).$$

There are three equivalent definitions of k-formal-schemes, all of them very important:

b) Let A be a <u>profinite</u> k-ring, i.e. a topological k-ring whose topology has a basis of neighbourhoods of zero consisting of ideals of finite codimension; one also can say that A is the inverse limit (as a topological ring) of discrete quotients which are finite k-rings. If $R \in \underline{Mf_k E}$, we define Spf (A)(R) as the set of all continuous homomorphisms of the topological k-ring A to the discrete k-ring R; if (A_i) is the family of discrete finite quotients of A defining its topology, then obviously $\mathrm{Spf}\ (A)(R) = \varinjlim\ \mathrm{Spf}\ (A_i)(R)$, and Spf A is a k-formal-scheme.

If $\varphi : A \longrightarrow B$ is a morphism of profinite k-rings, then $\mathrm{Spf}\ \varphi : \mathrm{Spf}\ B \longrightarrow \mathrm{Spf}\ A$ is obviously defined. We have then the

<u>Theorem</u>. <u>$A \longrightarrow \mathrm{Spf}\ A$ is an anti-equivalence of the category $\underline{PM_k}$ of profinite k-rings with the category of k-formal-schemes</u>.

<u>Proof</u>. We first prove that Spf is fully faithful: let A and B be two profinite k-rings and (A_i) be the family of all finite discrete quotients of A. We have isomorphisms

$$\underline{Mf_k E}(\mathrm{Spf}\ A, \mathrm{Spf}\ B) \simeq \varprojlim\ \mathrm{Spf}\ B(A_i) \simeq \varprojlim \underline{PM_k}(B, A_i) \simeq \underline{PM_k}(B, A).$$

We now prove that any k-formal-scheme F is isomorphic to a Spf A. By definition there is a directed projective system (A_i) of $\underline{Mf_k}$ such that F is isomorphic to $\varinjlim\ \mathrm{Spf}\ A_i$; let A be the topological k-ring $\varprojlim\ A_i$; we shall prove that A is a profinite k-ring and that $\varinjlim \mathrm{Spf}\ A_i \approx \mathrm{Spf}\ A$.

Let us fix an i; the images of the transition maps $f_{ij} : A_j \longrightarrow A_i,\ j \geqslant i$, form a directed decreasing set of sub-k-rings in the <u>finite</u> k-ring A_i; it follows that there is a $j(i) \geqslant i$ such that

$$f_{ij(i)}(A_{j(i)}) = \bigcap_{j \geqslant i} A_{ij};$$

it implies that, if we replace each A_i by $A'_i = \bigcap_{j \geqslant i} A_{ij}$, we change neither the topological k-ring A, nor the functor $\varinjlim \operatorname{Spf} A_i$. We can hence suppose that all transition maps $A_j \longrightarrow A_i$ are surjective. It is now sufficient to prove that the projections $A \longrightarrow A_i$ are *surjective*; this would imply both our assertions

Let now C_i be the k-vector space dual to A_i; the C_i form a directed inductive system with injective transition maps; call $C = \varinjlim C_i$; each canonical map $C_i \longrightarrow C$ is injective. Let C^* be the dual space of C. The dual maps $C^* \longrightarrow A_i$ are surjective and form a projective system; they factorize through A and the projections $A \longrightarrow A_i$ are *a fortiori* surjective. In fact, *the canonical map* $C^* \longrightarrow A$ *is bijective*; if $v \in C^*$ maps to zero on each A_i; then the linear form v over C vanishes over each C_i, hence is zero; conversely, if $a \in A$, then the projection of a on each A_i is a k-linear form on C_i; these linear forms match together, and define a k-linear form on C, which means that a belongs to the image of C^*

c) A k-*cogebra* is a k-vector space C together with a k-linear map $\Delta : C \longrightarrow C \otimes_k C$. We say that C is a k-*coring* if Δ is coassociative, cocommutative, and has a counit ε ; let us make these three notions precise.

1) Δ is *coassociative* if $(\Delta \otimes 1_C) \circ \Delta = (1_C \otimes \Delta) \circ \Delta$, in the following diagram

$$C \longrightarrow C \otimes C \underset{\Delta \otimes 1_C}{\overset{1_C \otimes \Delta}{\rightrightarrows}} C \otimes C \otimes C,$$

2) Δ is *cocommutative* if the image of Δ consists of symmetric tensors; equivalently, if $\sigma \circ \Delta = \Delta$ where $\sigma(x \otimes y) = y \otimes x$.

3) A <u>counit</u> ε to Δ is a k-linear form $\varepsilon : C \longrightarrow k$ such that the two maps

$$C \xrightarrow{\Delta} C \otimes C \xrightarrow{1_C \otimes \varepsilon} C \otimes k \xrightarrow{\simeq} C$$

$$C \xrightarrow{\Delta} C \otimes C \xrightarrow{\varepsilon \otimes 1_C} k \otimes C \xrightarrow{\simeq} C$$

are 1_C.

If C is a k-cogebra, then the dual k-vector space C^* has an algebra structure defined by $\langle x.y,u\rangle = \langle x \otimes y, \Delta u\rangle$, $x,y \in C^*$, $u \in C$. If C is a k-coring, then C^* is a ring.

Conversely, if A is a <u>finite</u> k-algebra, the dual space A^* has a natural cogebra structure, which is a coring structure if A is a ring. (If A is not finite, the dual space of $A \otimes A$ is <u>not</u> $A^* \otimes A^*$.).

The morphisms of k-corings are defined in an obvious way, and the k-corings form a category.

Let A and R be two finite k-rings, and A^* the dual k-coring of A. Linear maps $A \longrightarrow R$ correspond bijectively to elements of the tensor product $A^* \otimes R$; the k-linear maps Δ_{A^*} and ε_{A^*} extend to R-linear maps $A^* \otimes R \longrightarrow (A^* \otimes R) \otimes_R (A^* \otimes R)$ and $A^* \otimes R \longrightarrow R$ which also we denote by Δ and ε. We then have the easy

<u>Lemma</u>. <u>The k-linear map $A \longrightarrow R$ associated to $u \in A^* \otimes R$ is a ring homomorphism if and only if $\Delta u = u \otimes u$ and $\varepsilon u = 1$.</u>

We therefore have a functorial isomorphism

$$\text{Sp } A(R) = \{u \in A^* \otimes R \mid \Delta u = u \otimes u,\ \varepsilon u = 1\}.$$

For any k-coring C, we <u>define</u> the k-formal functor Sp^*C by $\text{Sp}^*C(R) = \{u \in C \otimes R \mid \Delta u = u \otimes u,\ \varepsilon u = 1\}$. We thus have a covariant functor Sp^* from the category of k-corings to the category of k-formal functors.

Theorem. The functor Sp^* is an equivalence between the category of k-corings and the category of k-formal-schemes.

Proof. As we have already seen Sp^* induces an equivalence between the category of finite k-corings and the category of finite k-schemes by the formula.

$$\operatorname{Spf} A = Sp^* A^*, \quad A \in \underline{M}f_k.$$

We have already seen that any k-formal-scheme F is an inductive limit of finite schemes $\operatorname{Spf}(A_i)$, with surjective transition maps $A_j \to A_i$; the inductive limit $C = \varinjlim A_i^*$ is naturally endowed with a k-coring structure, and, for any $R \in \underline{M}f_k$, we have

$$Sp^* C(R) \simeq \varinjlim Sp^* A_i^*(R) \simeq \varinjlim \operatorname{Spf} A_i(R) = F(R).$$

The only point that remains to be checked is that any k-coring is a union of finite dimensional ones:

Lemma. If C is a k-coring, and E a finite dimensional subvector space of C, there exists a finite-dimensional subvector space F of C with $E \subset F$ and $\Delta F \subset F \otimes F$.

We need only prove the lemma for $[E:k] = 1$, say $E = kx$. Let a_i be a k-basis of C and write $\Delta x = \sum x_i \otimes a_i$; put $F = \sum k x_i$; one has $x = (1 \otimes \varepsilon)\Delta(x) = \sum x_i \varepsilon(a_i) \in F$, and

$$\sum \Delta x_i \otimes a_i = (\Delta \otimes 1)\Delta x = (1 \otimes \Delta)\Delta x = \sum x_i \otimes \Delta a_i;$$

if $\Delta a_i = \sum b_{ij} \otimes a_j$, this gives $\Delta x_i = \sum x_i \otimes b_{ji} \in F \otimes C$, hence $\Delta F \subset F \otimes C$. Since Δ is cocommutative, we have $\Delta F \subset C \otimes F$, hence $\Delta F \subset F \otimes F$.

If C is a k-coring, let C^* be the k-dual space of C with the linear topology defined by the subspaces of C which are orthogonal to the finite-dimensional subcorings of C. Then, what we have proved already in b) gives: the k-ring C^* is profinite and

$$Sp^* C = \operatorname{Spf} C^*.$$

Conversely, we can recover C as the set of continuous linear forms on C^*: if A is a profinite k-ring, write A' for the set of continuous linear forms on A, then

$$\mathrm{Spf}\ A = \mathrm{Sp}^* A'.$$

d) The fourth definition of k-formal schemes is from a purely functorial point of view:

Theorem. A k-formal functor $\underline{Mf}_k \longrightarrow \underline{E}$ is a k-formal scheme if and only if it is a left exact functor.

We recall that a left exact functor is one which commutes with finite projective limits (i.e. which commutes with fibre products and with the final objects). Any Spf (A), $A \in \underline{Mf}_k$ is clearly left exact (this is true in any category, and is the very definition of finite projective limits) hence also any inductive limit of Spf (A_i), $A_i \in \underline{Mf}_k$, i.e. any k-formal-scheme, is left exact.

A proof of the converse can be found in D.G. V, § 2,3.1. This fourth definition will not be used in the sequel.

7. Operations on formal schemes.

A finite projective limit of k-formal-schemes is a k-formal-scheme. For instance let $F_1 \longrightarrow F \longleftarrow F_2$ be a diagram of k-formal-schemes corresponding to a diagram $A_1 \longleftarrow A \longrightarrow A_2$ of profinite k-rings; then $F_1 \times_F F_2$ is a k-formal scheme corresponding to the profinite k-ring $A_1 \widehat{\otimes}_A A_2$ where

$$A_1 \widehat{\otimes}_A A_2 = \varprojlim A_1/I_1 \otimes_A A_2/I_2$$

where I_1 (resp I_2) runs through the open ideals of A_1 (resp A_2) defining its topology; $A_1 \widehat{\otimes}_A A_2$ can also be defined as the completed ring of the usual tensor product $A_1 \otimes_A A_2$ for the topology given by the $A_1 \otimes I_2 + I_1 \otimes A_2$. The description from the coring point of view is a bit more difficult. Let $C_1 \xrightarrow{\varphi_1} C \xleftarrow{\varphi_2} C_2$ be the corresponding coring diagram. Then the k-coring D defining the fibre product is the kernel of the map from $C_1 \otimes C_2$ to C which sends $x_1 \otimes x_2$ to $\varphi_1(x_1)\,\varepsilon_2(x_2) - \varepsilon_1(x_1)\,\varphi_2(x_2)$; the canonical maps $D \longrightarrow C_1$ and $D \longrightarrow C_2$ are defined by $x_1 \otimes x_2 \longmapsto x_1\,\varepsilon_2(x_2)$ and $x_1 \otimes x_2 \longmapsto \varepsilon_1(x_1)x_2$.

More particularly $F_1 \times F_2$ corresponds to the profinite k-ring $A_1 \hat{\otimes} A_2$ and to the k-coring $A_1^* \otimes A_2^*$:

$$\mathrm{Spf}\,A_1 \times \mathrm{Spf}\,A_2 = \mathrm{Spf}\,(A_1 \hat{\otimes} A_2),$$

$$\mathrm{Sp}^* C_1 \times \mathrm{Sp}^* C_2 = \mathrm{Sp}^*(C_1 \otimes C_2),$$

(note that the maps $C_1 \otimes C_2 \longrightarrow C_i$, $i = 1,2$, are defined by the counits).

We shall need later the following lemma;

<u>Lemma</u>. Let $f = \mathrm{Spf}\,\psi = \mathrm{Sp}^*\varphi$ <u>be a morphism of</u> k-<u>formal schemes</u>. <u>Then</u> $\big($f <u>is a monomorphism</u>$\big) \Longleftrightarrow \big(\psi$ <u>is surjective</u>$\big) \Longleftrightarrow \big(\varphi$ <u>is injective</u>$\big)$.

Clearly, $(\varphi$ is injective$) \Longrightarrow (\psi$ is surjective$) \Longrightarrow (f$ is a monomorphism$)$. Conversely, if $f: X \longrightarrow Y$ is a monomorphism, then (general nonsense) the diagonal morphism $X \longrightarrow X \times_Y X$ is an isomorphism. If $\varphi: C \longrightarrow D$ is the corresponding coring morphism, then the following sequence

$$0 \longrightarrow C \xrightarrow{u} C \otimes C \xrightarrow{v} D$$

is exact, where $u(x) = x \otimes x$, $v(x \otimes y) = \varepsilon_C(x)\varphi(y) - \varepsilon_C(y)\varphi(x)$. If $\alpha \in \mathrm{Ker}\,\varphi$, then $\varepsilon_C(\alpha) = \varepsilon_D(\varphi(\alpha)) = 0$; it follows that for any $x \in C$, one has $v(x \otimes \alpha) = 0$; hence $C \otimes (\mathrm{Ker}\,\varphi) \subset u(C)$. This implies $\mathrm{Ker}\,\varphi = 0$, or $[C:k] = 1$, $\varphi = 0$; in the latter case, one has $\varepsilon_C = \varphi \circ \varepsilon_D = 0$, and this implies $C = 0$ (for instance because $1_C^* = 0$ implies $C^* = 0$).

The category of k-formal-schemes has <u>infinite direct sums</u>:

$$\coprod \mathrm{Spf}\,A_i = \mathrm{Spf}\prod A_i\ ;$$

$$\coprod \mathrm{Sp}^* C_i = \mathrm{Sp}^* \sum C_i\ .$$

A formal scheme F is said to be <u>local</u> if it is isomorphic to a $\mathrm{Spf}\, A$ where A is a local ring; equivalently, Card $F(K)$ must be 1 for all fields $K \in \underline{Mf}_k$. <u>Any formal scheme is a direct sum of local formal-schemes</u>: if $A = \varprojlim A/I_i$ is a profinite k-ring, let Ω be the set of all open maximal ideals of A; the artinian k-ring A/I_i is a product of local rings, which are the localized rings $(A/I_i)_{m/I_i}$ where m runs through the elements of Ω containing I_i; since $(A/I_i)_m = (A/I_i)_{m/I_i}$ if $m \supset I_i$ and $\{0\}$ otherwise, we have $A/I_i = \prod_{m\in\Omega} (A/I_i)_m$; defining A_m as the limit of the $(A/I_i)_m$, we get

$$A = \prod_{m\in\Omega} A_m$$

(each A_m being <u>local</u>, as a directed projective limit of local rings).

Let k' be an extension of k; we define the <u>base-change</u> functor by the following formulas

$$(\mathrm{Spf}\, A)\otimes_k k' = \mathrm{Spf}(A\otimes_k k'),$$

$$(\mathrm{Sp}^*C)\otimes_k k' = \mathrm{Sp}^*(C\otimes_k k').$$

If k'/k is finite, then this base-change functor is the obvious one, defined by $(F\otimes_k k')(R) = F(R_{[k]})$.

<u>If X is a k-scheme, then its completion $\hat{X}$ is a k-formal scheme</u>: more precisely, $\hat{X}$ is the direct sum of the $\mathrm{Spf}\, \hat{\underline{O}}_{X,x}$ where x runs through the points of X such that $[\mathcal{K}(x):k] < \infty$, and where $\hat{\underline{O}}_{X,x}$ is the completion of $\underline{O}_{X,x}$ for the topology defined by the ideals of finite codimension. If X is a (locally) algebraic k-scheme, then these x are precisely the closed point of X, and $\hat{\underline{O}}_{X,x}$ is the completion of $\underline{O}_{X,x}$ for the usual adic topology. For instance, if $X = \mathrm{Sp}\, A$, where A is a finitely generated k-ring, then $\hat{X} = \coprod \mathrm{Spf}\, \hat{A}_m$, where m runs through all maximal ideals of A, and $\hat{A}_m$ is the completion of the local ring A_m for the m-adic topology. The functor $X \mapsto \hat{X}$ is left exact and commutes with base-change.

8. Constant and etale schemes.

For the moment, let us drop the assumption that k is a field. Given a set E, we define the constant scheme E_k to be the direct sum (in the category of k-schemes)

$$E_k = (\mathrm{Sp}_k k)^{(E)};$$

equivalently, $|E_k|$ is the direct sum $(\mathrm{Spec}\ k)^{(E)}$. For any scheme X, we have canonical bijections

$$\underline{M}_k\underline{E}(E_k, X) \simeq \underline{M}_k\underline{E}(\mathrm{Sp}_k k, X)^{(E)} \simeq X(k)^{(E)} = \underline{E}(E, X(k)),$$

so that $E \longmapsto E_k$ is the right adjoint functor to $X \longmapsto X(k)$. This implies that $E \longmapsto E_k$ commutes with finite projective limits. If $k' \in \underline{M}_k$, one has a canonical isomorphism

$$E_{k'} \simeq E_k \otimes_k k' .$$

If X is a scheme, then $\underline{M}_k\underline{E}(X, E_k)$ can be identified with the set of continuous (i.e. locally constant) maps of $|X|$ to the discrete space E.

If E is finite, then E_k is affine and $O_k(E_k)$ is the k-ring k^E.

Let now k be again a field. We define the constant formal-scheme $\hat{E}_k$ as the completion of E_k, or equivalently, as the direct sum $(\mathrm{Spf}\ k)^{(E)}$. Then $\hat{E}_k \simeq \mathrm{Spf}\ k^E$, where k^E has the product topology.

A k-scheme (resp k formal-scheme) is called constant if it is isomorphic to an E_k (resp $\hat{E}_k$). The completion functor induces an equivalence between the category of constant k-schemes and the category of constant k-formal schemes.

We define now an etale k-scheme (resp an etale k-formal-scheme) to be a direct sum of Sp (resp Spf) of finite separable extensions of k. Let $\bar{k}$ be an algebraic closure of k, and k_s the subextension consisting of all separable elements of $\bar{k}$. Then:

__Proposition.__ __For a k-scheme X (resp. a k-formal scheme X), the following conditions are equivalent:__

X __is etale,__ $X\otimes_k \bar{k}$ __is constant,__ $X\otimes_k k_s$ __is constant.__

This proposition is an easy consequence of the following: if A is a k-ring, then A is a finite product of finite separable extensions of k if and only if $A\otimes_k \bar{k}$ is a finite power of k, or $A\otimes_k k_s$ a finite power of k_s.

Let Π be the Galois group of k_s/k; it is a profinite topological group. Let X be an etale k-scheme; then Π operates on the set $X(k_s)$ and the isotropy group of any $x\in X(k_s)$ is open in Π (one calls $X(k_s)$ a Π-set). The fundamental theorem of Galois theory is equivalent to:

__Proposition.__ $X \longrightarrow X(k_s)$ __is an equivalence between the category of etale k-schemes and the category of__ Π__-sets.__

Note also that $X \longmapsto \hat{X}$ is an equivalence between the categories of etale k-schemes and etale k-formal schemes.

9. __The Frobenius morphism.__

We suppose now that the characteristic p of the field k is >0. For any k-ring A, we denote $f_A : A \longrightarrow A$ the map $x \longmapsto x^p$; we denote by $A_{[f]}$ the k-ring deduced from A by the scalar restriction $f_k : k \longrightarrow k$, and $A^{(p)} = A\otimes_{k,f_k} k$ the k-ring obtained by the scalar extension f_k.

Then $f_A : A \longrightarrow A_{[f]}$ is a k-ring morphism, and defines a k-ring morphism

$$F_A : A^{(p)} \longrightarrow A, \quad F_A(x\otimes\lambda) = x^p\lambda .$$

If X a k-functor, we put $X^{(p)} = X\otimes_{k,f} k$, so that

$$X^{(p)}(R) = X(R_{[f]});$$

and we define the <u>Frobenius morphism</u> $F_X : X \longrightarrow X^{(p)}$ by

$$F_X(R) = X(f_R) : X(R) \longrightarrow X^{(p)}(R) = X(R_{[f]}).$$

For example, if $X = Sp_k A$, then $X^{(p)} = Sp_k A^{(p)}$ and $F_X = Sp_k F_A$. More generally, if X is a k-scheme, $X^{(p)}$ is a k-scheme. If $k = \mathbb{F}_p$, then $X^{(p)} = X$, but $F_X \neq id_X$ in general. If k' is an extension of k, then $(X \otimes_k k')^{(p)} = X^{(p)} \otimes_k k'$ and $F_{X \otimes_k k'} = F_X \otimes_k k'$ (obvious from the definitions).

Analogous definitions can be given for formal-functors and formal-schemes and the completion functor commutes with these constructions.

<u>Proposition</u>. <u>Let</u> X <u>be a</u> k-<u>formal scheme, or a locally algebraic</u> k-<u>scheme; then</u> X <u>is etale if and only if</u> F_X <u>is a monomorphism, or if and only if</u> F_X <u>is an isomorphism</u>.

Let us give the proof in the case of a locally algebraic k-scheme. We can replace X by $X \otimes_k \bar{k}$, hence suppose $k = \bar{k}$. If X is constant, then F_X is an isomorphism. Conversely, suppose F_X is a monomorphism; let $U = Sp\ A$ be an algebraic open affine subscheme of X; then F_U is a monomorphism and we have to prove that A is a finite power of k. Let m be a maximal ideal of A; write $A/m^2 = A/m \oplus m/m^2$ and look at the two following maps: the first one is the canonical map $u : A \longrightarrow A/m^2$, the second one is $v : A \longrightarrow A/m \longrightarrow A/m \oplus m/m^2$. Trivially $u \circ F_A = v \circ F_A$; but by hypothesis F_A is an epimorphism of $\underline{M}_k$, and this implies $u = v$ i.e. $m/m^2 = 0$. For any maximal ideal m of A, we therefore have $m = m^2$, and this in turn implies in a well-known manner that $A \xrightarrow{\sim} k^n$.

10. <u>Frobenius map and symmetric products</u>.

Suppose again $p \neq 0$. Let V be a k-vector space, $\otimes^p V$ the p-fold tensor power of V, $TS^p V$ the subspace of symmetric tensors and $s : \otimes^p V \longrightarrow TS^p V$ the <u>symmetrization operator</u>: $s(a_1 \otimes \ldots \otimes a_p) = \sum a_{\sigma(1)} \otimes \cdots \otimes a_{\sigma(p)}$, where σ runs through the symmetric group $\mathfrak{S}_p$. Let $\alpha_V : V^{(p)} \longrightarrow TS^p V$ be the linear map sending $a \otimes \lambda$ to $\lambda \cdot (a \otimes \ldots \otimes a)$.

<u>Lemma</u>. <u>The composite map</u> $V^{(p)} \xrightarrow{\alpha_V} TS^pV \longrightarrow TS^pV/s(\otimes^pV)$ <u>is bijective</u>.

The proof is an easy exercise in linear algebra.

Define the canonical map $\lambda_V : TS^pV \longrightarrow V^{(p)}$ by $\lambda_V \circ s = 0$, $\lambda_V \circ \alpha_V = \mathrm{Id}$.

If A is a k-ring, then TS^pA is a ring and λ_A a k-ring homomorphism (because $s(\otimes^pA)$ is an ideal in TS^pA by the formula $s(uv) = us(v)$ for u symmetric). If $X = Sp\,A$, we denote $Sp(TS^pA)$ by S^pX (p-fold symmetric power of X) One has then the following <u>commutative diagram</u>

$$\begin{array}{ccc} X^p & \xrightarrow{\quad\text{can}\quad} & S^pX \\ \uparrow & & \uparrow{\scriptstyle Sp\,\lambda_A} \\ X & \xrightarrow{\quad F_X\quad} & X^{(p)} \end{array}$$

which gives another definition for F_X.

Let now C be a k-coring, and consider the Frobenius morphism $F : Sp^*C \longrightarrow Sp^*C^{(p)}$ (it is clear that $(Sp^*C)^{(p)} = Sp^*C^{(p)}$, where $C^{(p)} = C \otimes_{k,f} k$). There exists a unique coring map $V_C : C \longrightarrow C^{(p)}$ such that $F = Sp^*V_C$. The pth iterate $\Delta_p : C \rightarrow \otimes^pC$ of $\Delta : C \longrightarrow \otimes^2C$ (defined inductively by $\Delta_2 = \Delta$, $\Delta_3 = (1 \otimes \Delta) \circ \Delta = (\Delta \otimes 1) \circ \Delta, \dots$) maps C in TS^pC, and we have the

<u>Theorem</u>. $V_C : C \longrightarrow C^{(p)}$ <u>is the composite map</u> $C \xrightarrow{\Delta_p} TS^pC \xrightarrow{\lambda_C} C^{(p)}$.

<u>Proof</u>. Let A be the (profinite) k-ring C^*; then $A^{(p)} \simeq (C^{(p)})^* = (C^*)^{(p)}$. If $a \in A$, $x \in C$, one has by definition $\langle a \otimes 1, V(x)\rangle = \langle a^p, x\rangle$ where $a \otimes 1 \in (C^*)^{(p)} = C^* \otimes_{k,f} k$ and $V(x) \in C^{(p)}$. By definition of the multiplication of A, one also has $\langle a^p, x\rangle = \langle a \otimes \dots \otimes a, \Delta_p x\rangle$ in the duality between $\otimes^pA$ and $\otimes^pC$. But $a \otimes \dots \otimes a$ is symmetric, and $\Delta_p(x) = \alpha_C(y) + s(v)$ for $y = \lambda_C \Delta_p(x)$ and a suitable $v \in \otimes^pC$. Since $\langle a \otimes \dots \otimes a, s(v)\rangle = 0$, this gives

$$\langle a \otimes 1, V(x)\rangle = \langle a \otimes \dots \otimes a, \alpha_C(x)\rangle = \langle a \otimes 1, y\rangle$$

and $V(x) = y = \lambda_C \Delta_p(x)$, as claimed above.

Corollary. $X = Sp^*C = Spf\ A$ is etale if and only if F_A is surjective (resp. bijective) and if and only if V_C is injective (resp. bijective).

CHAPTER II

GROUP-SCHEMES AND FORMAL GROUP-SCHEMES

1. Group-functors.

Let k be a ring. A group law on a k-functor $G \in \underline{M}_k$ is a family of group-laws on all the $G(R)$, $R \in \underline{M}_k$, such that each functoriality map $G(R) \longrightarrow G(S)$ is a homomorphism. Equivalently, a group law on G is a morphism

$$\pi : G \times G \longrightarrow G$$

such that

$$\pi(R) : G(R) \times G(R) \longrightarrow G(R)$$

is a group law for all R; this condition is equivalent to the axioms (Ass), (Un), (Inv):

(Ass) The two morphisms $\pi \circ (\pi \times 1_G)$ and $\pi \circ (1_G \times \pi)$ from $G \times G \times G$ to G are equal.

(Un) There exists an element $1 \in G(k)$ (or equivalently a morphism $e : \mathrm{Sp}\, k \longrightarrow G$) such that $\pi \circ (1_G \times e)$ and $\pi \circ (e \times 1_G)$ are equal to 1_G.

(Inv) There exists a morphism $\sigma : G \longrightarrow G$ such that the two morphisms

$G \xrightarrow{(1_G, \sigma)} G \times G \xrightarrow{\pi} G$ and $G \xrightarrow{(\sigma, 1_G)} G \times G \xrightarrow{\pi} G$ are equal to 1_G.

We are principally interested in commutative group laws, i.e. such that $G(R)$ is commutative for all R, i.e.

(Com) If $\tau : G \times G \longrightarrow G \times G$ is the symmetry, then $\tau \circ \pi = \pi$.

A k-group-functor is a pair (G, π) where G is a k-functor and π a group-law on G. The k-group functors form a category, a homomorphism $f : G \longrightarrow H$ being a morphism such that $f(R) : G(R) \longrightarrow H(R)$ is a group-homomorphism for each R, or equivalently such that $(f\times f)\circ\Delta_G = \Delta_H \circ f$. The category $\underline{Gr}_k$ of k-group-functors has projective limits. For instance:

— The final object $e_k = \mathrm{Sp}\, k$ has a unique group law.

— If $G \longrightarrow H \longleftarrow K$ is a diagram of $\underline{Gr}_k$, the fibre product $G\times_H K$ has an obvious group law, for which it is the fibre product in $\underline{Gr}_k$.

— In particular, if $f: G \longrightarrow H$ is a homomorphism, the kernel Ker f of f is the sub-functor $G\times_H e_k$ of G; equivalently

$$(\mathrm{Ker}\, f)(R) = \mathrm{Ker}(f(R): G(R) \longrightarrow H(R)).$$

The homomorphism f is a monomorphism if and only if $\mathrm{Ker}\, f = e_k$.

— The definition of a subgroup functor is clear.

A k-group-scheme or k-group is a k-group functor whose underlying k-functor is a scheme.

The base-change functor $\underline{Gr}_k \longrightarrow \underline{Gr}_{k'}$, for $k' \in M_k$ is obviously defined.

2. Constant and etale k-groups.

The functor $E \longrightarrow E_k$ from sets to k-schemes commutes with products and final objects; it follows that E_k has a natural group-law if E is a group. Such a k-group is called a constant k-group. Suppose k is a field and Π the Galois group of k_s/k; the functor $X \longrightarrow X(k_s)$ from etale k-schemes to Π-sets is an equivalence (I.8); it follows then from the definition of a k-group, and the fact that a product of etale schemes is also etale:

Proposition. The functor $X \longrightarrow X(k_s)$ is an equivalence between the category of etale k-groups (resp. commutative etale k-groups) and the category of Π-groups (resp. commutative Π-groups = Galois modules over Π).

Moreover, X is an etale k-group if and only if $X \otimes_k k_s$ is a constant k-group.

3. Affine k-groups.

Let $G = Sp_k A$ be an affine k-scheme. The morphisms $\pi : G\times G \longrightarrow G$ are the $Sp_k \Delta$ where $\Delta : A \longrightarrow A \otimes_k A$ is a k-ring morphism. Moreover π satisfies Ass, Com, Un if and only if Δ is coassociative, cocommutative, has a counit. The condition (Inv) is equivalent to (Coinv): there exists $\sigma : A \longrightarrow A$ such that the composite maps

$$A \xrightarrow{\Delta} A \otimes A \xrightarrow{1\otimes\sigma} A \otimes A \xrightarrow{\text{product}} A$$

$$A \xrightarrow{\Delta} A \otimes A \xrightarrow{\sigma\otimes 1} A \otimes A \xrightarrow{\text{product}} A$$

are the composite map $A \xrightarrow{\varepsilon} k \longrightarrow A$.

Such a σ is called an involution, or antipodism. If one identifies A with $O(G)$, $A \otimes A$ with $O(G\times G)$, then

$$(\Delta f)(x,y) = f(xy), \quad \sigma f(x) = f(x^{-1}), \quad \varepsilon f = f(1),$$

for $x,y \in G(R)$, $R \in \underline{M}_k$.

We shall be interested in commutative groups. Let us define a k-biring A as a k-module, together with a structure of k-ring and a structure of k-coring, which are compatible in either of the two equivalent following senses:

— the product $A\otimes A \longrightarrow A$ is a k-coring morphism.

— the coproduct $A \longrightarrow A \otimes A$ is a k-ring morphism.

Then, the category of commutative affine k-groups is antiequivalent to the category of k-birings with antipodism by $G \longmapsto O(G)$ and $A \longmapsto Sp\ A$ (the morphisms of birings are defined in the obvious way).

A very useful remark is the following: let G be an affine k-group and $A = O(G)$ [then $\underline{M}_k(A,R) \simeq G(R)$ for any $R \in \underline{M}_k$],

1) in the group $G(A \otimes A) = \underline{M}_k(A, A \otimes A)$, Δ_A is the product of the two canonical maps $i_1 : a \longmapsto 1 \otimes a$ and $i_2 : a \longmapsto a \otimes 1$,

2) in the group $G(A) = \underline{M}_k(A,A)$, σ_A is the inverse of 1_A,

3) ε_A is the identity of $G(k) = \underline{M}_k(A,k)$.

These facts are trivial: for instance 1) says that if H is a group, the map $(x,y) \longrightarrow xy$ is the product of $(x,y) \longrightarrow x$ and $(x,y) \longrightarrow y$.

Example 1. The additive group $\underline{\alpha}_k$ is defined as follows: $\underline{\alpha}_k(R)$ is the additive group of R; then, by the above remarks:

$$O(\underline{\alpha}_k) = k[T]$$

(T is the identity $\underline{\alpha}_k \longrightarrow \underline{O}_k$), $\Delta T = T \otimes 1 + 1 \otimes T$, $\sigma T = -T$, $\varepsilon T = 0$.

Example 2. The multiplicative group $\underline{\mu}_k$ is defined as follows: $\underline{\mu}_k(R)$ is the multiplicative group of invertible elements of R; hence

$$O(\underline{\mu}_k) = k[T, T^{-1}]$$

($T : \underline{\mu}_k \longrightarrow \underline{O}_k$ is the inclusion), $\Delta T = T \otimes T$, $\sigma T = T^{-1}$, $\varepsilon T = 1$.

Example 3. Let $n \geqslant 1$ be an integer. We define a group homomorphism $\underline{\mu}_k \xrightarrow{n} \underline{\mu}_k$ by $x \longmapsto x^n$. The kernel of this homomorphism is denoted by ${}_n\underline{\mu}_k$. Hence

$${}_n\underline{\mu}_k(R) = \{x \in R,\ x^n = 1\}$$

$$O({}_n\underline{\mu}_k) = k[T]/T^n - 1,$$

with the same formulas as above.

Note that, if k is a field and n is not 0 in k, ${}_n\mu_k$ is <u>etale</u> (because $T^n - 1$ is a separable polynomial) and ${}_n\mu_k(k_s)$ is the Galois module of nth roots of unity.

<u>Example 4</u>. Let k be a field with characteristic $p \neq 0$. One defines ${}_{p^r}\alpha_k$ as the kernel of the homomorphism $x \mapsto x^{p^r}$ of α_k in itself. Hence

$$ {}_{p^r}\alpha_k(R) = \{x \in R,\ x^{p^r} = 0\}, $$

$$ O({}_{p^r}\alpha_k) = k[T]/T^{p^r}. $$

Note that ${}_{p^r}\alpha_k(K) = \{0\}$ for any <u>field</u> K.

Remark that $\alpha_k \otimes_k k' = \alpha_{k'}$, $\mu_k \otimes_k k' = \mu_{k'}$, ...

The remarks we made about the construction of Δ, ε can be generalized in the following way. Let H be any k-group functor, and $G = Sp_k A$ be an affine k-group. Let $f \in \underline{M}_k\underline{E}(G,H) \simeq H(A)$; consider the three maps $i_1, i_2, \Delta : A \longrightarrow A \otimes A$. Then:

<u>Lemma</u>. <u>The element</u> $f \in H(A)$ <u>is a group homomorphism from</u> G <u>to</u> H <u>if and only if in the group</u> $H(A \otimes A)$, <u>one has</u> $\Delta(f) = i_1(f) i_2(f)$. Because, if $H(A \otimes A)$ is identified with $\underline{M}_k\underline{E}(G \times G, H)$, then $\Delta(f)$, $i_1(f)$ and $i_2(f)$ map (x,y) to $f(xy)$, $f(x)$, $f(y)$ respectively.

<u>Examples</u>. $\underline{Gr}_k(G, \alpha_k) = \{x \in A,\ \Delta x = x \otimes 1 + 1 \otimes x\}$,

$$ \underline{Gr}_k(G, {}_{p^r}\alpha_k) = \{x \in A,\ x^{p^r} = 0,\ \Delta x = x \otimes 1 + 1 \otimes x\}, $$

$$ \underline{Gr}_k(G, \mu_k) = \{x \in A, \Delta x = x \otimes x,\ \varepsilon x = 1\}. $$

As for the latter, remark that the lemma gives: $x \in A = \underline{M}_k\underline{E}(G,\underline{O}_k)$ is a homomorphism from G to μ_k if and only if $\Delta x = x \otimes x$, <u>and</u> x <u>is invertible</u>. But this implies $\varepsilon x = 1$ (because a group homomorphism sends 1 to 1); conversely, if $\Delta x = x \otimes x$ and $\varepsilon x = 1$, then by (Coinv) $x\sigma(x) = \varepsilon x = 1$.

$$\underline{Gr}_k(G, {}_n\mu_k) = \{x \in A, x^n = 1, \Delta x = x \otimes x, \varepsilon x = 1\}.$$

4. k-<u>formal-groups</u>, <u>Cartier</u> <u>duality</u>.

Suppose now that k is a field. The definitions of $n^o 1$ can be carried <u>mutatis</u> <u>mutandis</u> to k-formal functors. A k-<u>formal</u> <u>group</u> is a k-formal-group-functor whose underlying k-formal-functor is a k-formal-scheme. For k-formal groups, we can repeat $n^o 3$, replacing tensor products, by completed tensor products: the coproduct maps A to $A \hat{\otimes} A$, ... If G is a k-group, then $\hat{G}$ has a natural structure of a k-formal group. For instance, $G \longrightarrow \hat{G}$ is an equivalence between constant (resp. etale, resp. finite) k-groups and constant (resp. etale, resp. finite) k-formal groups.

It is more interesting to look at formal-groups from the point of view of k-corings. Let $G = Sp^*C$ be a k-formal-scheme; to give a morphism $\pi: G \times G \longrightarrow G$ is equivalent to give a k-coring map $C \otimes C \longrightarrow C$ i.e. an algebra structure on C compatible with the coring structure; moreover, π is a group law (resp. a commutative group law) if and only if this algebra structure is associative, has a unit element and an antipodism (same axiom as (Coinv)) (resp. and is commutative). In particular, $C \longrightarrow Sp^*C$ is an equivalence between k-<u>birings</u> <u>with</u> <u>antipodism</u> and <u>commutative</u> k-<u>formal-groups</u>. It follows that $Sp\ C \longrightarrow Sp^*C$ is <u>an</u> <u>antiequivalence</u> <u>between</u> <u>commutative</u> <u>affine</u> k-<u>groups</u> <u>and</u> <u>commutative</u> k-<u>formal-groups</u>. This can also be explained as follows:

For any commutative k-group-functor G, we define the <u>Cartier dual</u> of G as the commutative k-group-functor $D(G)$ such that, for $R \in \underline{M}_k$,

$$D(G)(R) = \underline{Gr}_R(G \otimes_k R, \mu_R);$$

if G and H are two commutative k-group-functors, then it is equivalent either to give a homomorphism $G \longrightarrow D(H)$, or a homomorphism $H \longrightarrow D(G)$, or a "bilinear" morphism $G \times H \longrightarrow \mu_k$. In particular, there is canonical <u>biduality homomorphism</u>

$$\alpha_G : G \longrightarrow D(D(G)).$$

If $k' \in \underline{M}_k$, then $D(G \otimes_k k') = D(G) \otimes_k k'$, and $\alpha_{G \otimes_k k'} = \alpha_G \otimes_k k'$.

<u>Theorem</u> 1) <u>If G is an affine commutative k-group, $\widehat{D(G)}$ is a commutative k-formal group. More precisely, if $G = Sp\, A$, where A is a k-biring with antipodism, then $\widehat{D(G)} = Sp^* A$. The functor $G \longrightarrow \widehat{D(G)}$ is an antiequivalence between affine commutative k-groups and commutative k-formal-groups.</u>

2) <u>If G is a finite commutative k-group, then $D(G)$ also is; α_G is an isomorphism, and $G \longrightarrow D(G)$ induces a duality in the category of finite commutative groups. Moreover $rk(G) = rk(D(G))$.</u>

Let $G = Sp\, A$, where A is a k-biring with involution. Then, for $R \in \underline{Mf}_k$

$$\widehat{D(G)}(R) = \underline{Gr}_R(G \otimes_k R, \mu_R) = \{x \in A \otimes_k R,\ \Delta x = x \otimes x,\ \varepsilon x = 1\} = Sp^* A(R);$$

to prove 1), it remains only to show that the multiplication in A giving the group structure of $D(G)$ is the given one; this verification is straightforward. The proof of 2) is similar.

<u>Examples</u> 1) $D((\mathbb{Z}/n\mathbb{Z})_k) = {}_n\mu_k$ and conversely (exercise).

2) (Charac (k) = p $\neq$ 0) There is a canonical bilinear morphism

$$f: {}_p\underline{\alpha}_k \times {}_p\underline{\alpha}_k \longrightarrow \underline{\mu}_k$$

given by $f(x,y) = \exp(xy) = 1 + xy + \ldots + (xy)^{p-1}/(p-1)!$. It defines an isomorphism $D({}_p\underline{\alpha}_k) \simeq {}_p\underline{\alpha}_k$.

3) $D(\underline{\mu}_k) = \mathbb{Z}_k$, hence $\widehat{D(\underline{\mu}_k)} = \widehat{\mathbb{Z}_k}$ (exercise).

5. <u>The Frobenius and the Verschiebung morphisms</u>.

Suppose charac (k) = p $\neq$ 0. The functors $G \to G^{(p)}$ and the morphism $F_G: G \longrightarrow G^{(p)}$ commute with products. This implies that, if G is a k-group-functor, then $G^{(p)}$ has a natural structure of a k-group-functor, and F_G is a homomorphism. The same is true for k-formal-group-functors.

We define $G^{(p^n)}$ by $G^{(p^n)} = (G^{(p^{n-1})})^{(p)}$, and $F^n_G: G \longrightarrow G^{(p^n)}$ by $F^n_G = F^{n-1}_{G^{(p)}} \circ F_G$.

Let G be a commutative affine k-group. We have $D(G^{(p)}) = D(G)^{(p)}$. By Cartier duality, there is therefore a unique homomorphism (<u>the Verschiebung morphism</u>)

$$V_G: G^{(p)} \longrightarrow G$$

such that $\widehat{D}(V_G) = F_{\widehat{D}(G)}$. If G = Sp A, then $\widehat{D(G)} = Sp^* A$, and we see that V_G = Sp V_A (V_A has been defined in I, n°10).

In the same way, we define the Verschiebung homomorphism for <u>commutative</u> k-<u>formal</u> groups. One defines also $V^n_G: G^{(p^n)} \longrightarrow G$ in the same way as F^n_G.

If $f: G \longrightarrow H$ is an homomorphism of commutative affine k-groups (or k-formal groups), then the following diagram is clearly commutative:

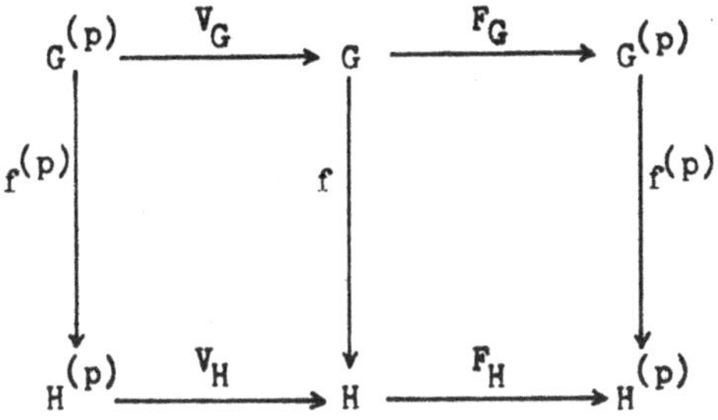

<u>Proposition</u>. <u>If</u> G <u>is an affine commutative k-group</u> (resp. <u>a commutative k-formal group</u>), <u>then</u>

$$V_G \circ F_G = p.\mathrm{id}_G, \quad F_G \circ V_G = p.\mathrm{id}_G(p).$$

Equivalently, $V_G(F_G(x)) = px$, $F_G(V_G(x)) = px$ (additive notation).

It is sufficient to prove this for the affine case, because the formal case follows by Cartier duality. Moreover, the first formula (for any G) implies the second one: by the functoriality of F and V, one has a commutative diagram,

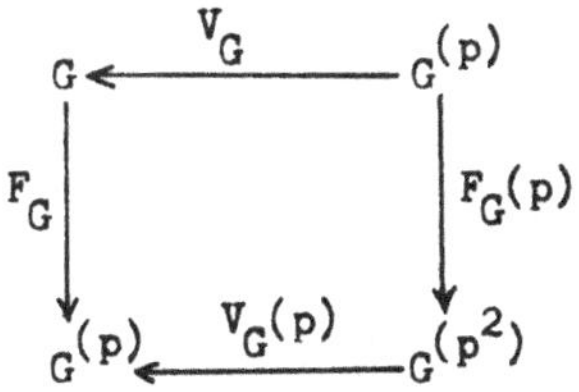

and $F_G \circ V_G = V_G(p) \circ F_G(p)$.

To prove $V_G \circ F_G = p\ \mathrm{id}_G$, we use I, n°10. One has a commutative diagram (where $A = O(G)$):

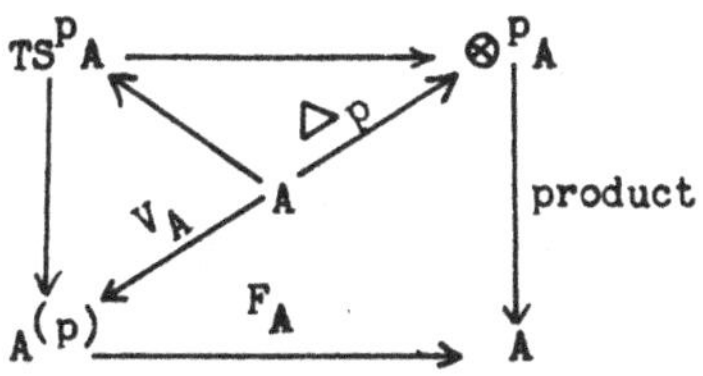

or

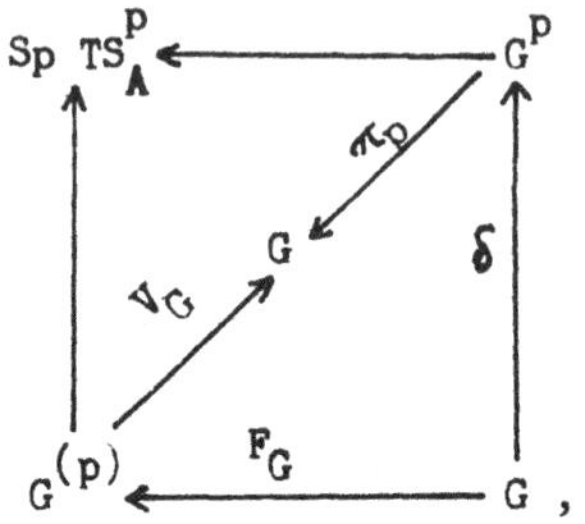

with $\delta(g) = (g,\ldots,g)$, and $\pi_p(g_1 \ldots g_p) = g_1 + \ldots + g_p$. Then $V_G \circ F_G = \pi_p \delta = p\, id_G$.

Remark. The above diagram gives a direct definition of V_G.

Examples. $V: \mu_k \longrightarrow \mu_k$ is the identity, $V: \alpha_k \longrightarrow \alpha_k$ is zero. This follows from the facts that F is an epimorphism for α_k and μ_k and that $p\, id_{\mu_k} = F_{\mu_k}$, $p\, id_{\alpha_k} = 0$.

6. The category of affine k-groups.

Recall that k is supposed to be a field. Let $\underline{AC}_k$ be the category of all affine commutative k-groups.

Theorem 1. (Grothendieck): The category $\underline{AC}_k$ is abelian.

a) $\underline{AC}_k$ is an additive category: Clear.

b) Any morphism $f: G \longrightarrow H$ of $\underline{AC}_k$ has a kernel: one has

$$\mathrm{Ker}\, f = G \times_H e\ , \mathcal{O}(\mathrm{Ker}\, f) = \mathcal{O}(G)/\mathfrak{m}(H)\mathcal{O}(G)$$

($m(H) = \mathrm{Ker}\ \varepsilon_H : O(H) \longrightarrow k$). Remark that $O(G) \longrightarrow O(\mathrm{Ker}\ f)$ is surjective.

c) Any morphism $f: G \longrightarrow H$ of $\underline{AC_k}$ has a cokernel: One takes Coker f such that

$$O(\mathrm{Coker}\ f) = O(H)^G = \{f \in O(H),\ f(g+h) = f(h)\ \ \forall g \in G(R),\ h \in H(R)\}$$

$$= \{f \in O(H),\ (1 \otimes O(f))\Delta_H(f) = f \otimes 1\}.$$

Remark that $O(\mathrm{Coker}\ f) \longrightarrow O(H)$ is injective.

d) There is only one thing more to prove, and this is the fundamental fact, that any monomorphism is a kernel, and any epimorphism is a cokernel. More precisely

Theorem 2. *Let $f: G \longrightarrow H$ be a morphism of AC_k.*

1) The following conditions are equivalent: f is a monomorphism, $O(f)$ is surjective (i.e. G is a closed subgroup of H), f is a kernel.

2) The following conditions are equivalent: f is an epimorphism, $O(f)$ is injective, $O(f): O(H) \longrightarrow O(G)$ makes $O(G)$ a faithfully flat $O(H)$-module, $O(f)$ is a cokernel.

For a proof see D.G. III, 3.7.4. The main point is (f mono) $\Longrightarrow$ (f kernel) or equivalently (f mono) $\Longrightarrow$ ($f = \mathrm{Ker}(\mathrm{coker}\ f)$).

Corollary 1. *If k' is an extension of k, then $G \longmapsto G \otimes_k k'$ is an exact functor.*

Clear: It respects kernels and cokernels.

Corollary 2. *Let $0 \longrightarrow K \longrightarrow G \longrightarrow H \longrightarrow 0$ be an exact sequence, then the $O(G)$-algebra $O(G) \otimes_{O(H)} O(G)$ is isomorphic to $O(G) \otimes O(K)$.*

Clear: The morphism $(g,k) \longmapsto (g,gk)$ of $G \times K \longrightarrow G \times_H G$ is an isomorphism.

Corollary 3. *If* $0 \longrightarrow K \longrightarrow G \longrightarrow H \longrightarrow 0$ *is a exact sequence with* K *algebraic (resp. finite of rank* r*); then* $O(G)$ *is a finitely presented* $O(H)$*-ring (resp. a finitely generated projective* $O(H)$*-module of rank* r*).*

Because it becomes so after the faithfully flat scalar-extension $O(H) \longrightarrow O(G)$ (Corollary 2).

Corollary 4. *If* $0 \longrightarrow K \longrightarrow G \longrightarrow H \longrightarrow 0$ *is an exact sequence, then* G *is algebraic (resp. finite) if and only if* H *and* K *are. In the finite case, one has* $rk(G) = rk(K).rk(H)$.

If $O(G)$ is finitely generated or finite, so is the subalgebra $O(H)$ and the quotient $O(K)$. The converse and the last assertion follow from corollary 3.

Corollary 5. *If* $f: G \longrightarrow H$ *is an epimorphism (resp. and if* $\operatorname{Ker} f$ *is algebraic, resp. finite) and if* $R \in M_k$, *and* $h \in H(R)$, *there exists an* R*-ring* S *faithfully flat (resp. and finitely presented, resp. finite and projective) and a* $g \in G(S)$ *such that* $f(g) = h_S$

Clear from Corollary 4: h is given as a map $O(H) \longrightarrow R$; take $S = O(G) \otimes_{O(H)} R$.

Corollary 6. *If* $f: G \longrightarrow H$ *is an epimorphism with* $\operatorname{Ker} f$ *algebraic, if* $L \in M_k$ *is a field, and* $h \in H(L)$, *there exists a finite extension* L' *of* L *and a* $g \in G(L')$ *with* $f(g) = h_{L'}$.

Follows from corollary 5 by the Nullstellensatz.

Remark. If f is an epimorphism (*without any hypothesis* on $\operatorname{Ker} f$), then $f(L)$ is surjective for any *algebraically closed field* L (D.G. III, 3.7.6).

By Cartier Duality *the category of commutative* k*-formal-groups also is abelian*, and $\operatorname{Spf} \varphi$ is a monomorphism (resp. an epimorphism) if and only if φ is surjective (resp. injective).

Theorem 3. a) The abelian category AC_k satisfies the axiom (AB 5^*): it has directed projective limits, and a directed projective limit of epimorphisms is an epimorphism.

b) The artinian objects of AC_k are the algebraic groups. Any object of AC_k is the directed projective limit of its algebraic quotients.

a) is clear from Th.2: one has $\varprojlim \mathrm{Sp}\,\varphi_i = \mathrm{Sp}\,\varinjlim \varphi_i$ and a directed inductive limit of injective maps is injective.

b) see D.G. II, 2.3.7.

By Cartier duality, the dual statements hold for the category of commutative k-formal-groups.

From now on we shall mainly speak about commutative groups. We say group instead of commutative group unless otherwise stated. From now on also, k is a field, p denotes the characteristic of k, and $\pi = \mathrm{Gal}(k_s/k)$. Our main interest will be the case $p \neq 0$. As we shall see, the case $p = 0$ is rather trivial.

7. Etale and connected formal-groups.

We already defined and studied etale affine (resp. formal) groups. They are equivalent to finite (resp. all) Galois modules by

$$E \longrightarrow (E \otimes_k k_s)(k_s) = \bigcup_{\substack{K/k \text{ sep.} \\ \text{finite}}} E(K).$$

If $p \neq 0$, then G is etale iff $\mathrm{Ker}\, F_G = e$, and this implies that F is an isomorphism (I,9). It follows that subgroups, quotients and extensions (direct limits in the formal case) of etale groups also are etale. The same statement is true if $p = 0$.

Recall, that the formal-group $G = \mathrm{Spf}\ A$ is <u>local</u> (We shall also say <u>connected</u>) if A is local or equivalently if $G(K) = \{0\}$ for any field K. A morphism from a connected group to an etale group is zero.

<u>Proposition. Let G be a formal-group.</u>

a) <u>There is an exact sequence (unique up to isomorphism)</u>

$$0 \longrightarrow G^{o} \longrightarrow G \longrightarrow \pi_0(G) \longrightarrow 0$$

<u>where G^o is connected, and $\pi_0(G)$ etale. If $R \in \underline{Mf}_k$ and $\underline{n}$ is the nilradical of R then</u> $G^o(R) = \mathrm{Ker}(G(R) \longrightarrow G(R/\underline{n}))$. If $p \neq 0$, <u>then G^o is the limit of the</u> $\mathrm{Ker}(F^n_G : G \longrightarrow G^{(p^n)})$, $n \geq 0$. If $k \to k'$ <u>is an extension then</u> $(G \otimes_k k')^o = G^o \otimes_k k'$, $\pi_0(G \otimes_k k') = \pi_0(G) \otimes_k k'$.

b) <u>If k is perfect, there is a unique isomorphism</u> $G = G^o \times \pi_0(G)$.

<u>Proof</u>. Write $G = \mathrm{Spf}\ A = \coprod \mathrm{Spf}\ A_m$. Let A^o be the local factor A_{m_o} corresponding to the ideal $m_o = \mathrm{Ker}(\varepsilon : A \longrightarrow k)$. Call $G^o = \mathrm{Spf}\ A^o$; by construction, $G^o(R) = \mathrm{Ker}(G(R) \longrightarrow G(R/\underline{n}))$ for $R \in \underline{Mf}_k$; it follows that G^o is a subgroup of G. If $k \longrightarrow k'$ is an extension, then $A \otimes_k k'$ is local, because the residue field of A^o is k; it follows that $(G \otimes_k k')^o = G^o \otimes_k k'$. Suppose $p \neq 0$, then $\mathrm{Ker}\ F^n_G = \mathrm{Spf}\ A/m_o^{\{p^n\}}$, where $m_o^{\{p^n\}}$ is the closed ideal of A generated by the x^{p^n}, $x \in m_o$; hence $\bigcup_n \mathrm{Ker}\ F^n_G = \mathrm{Spf}(\varprojlim A/m_o^{\{p^n\}}) = \mathrm{Spf}\ A_o = G^o$. To prove a), it only remains to show that G/G^o is <u>etale</u>.

Remark first that G is etale if and only if $G^o = e$: replacing k by $\bar{k}$ we can suppose k to be algebraically closed; if $G^o = e$ then $A^o = k$; but then all the A_m are isomorphic (by translation); hence $A \simeq k^E$ and G is etale. To prove that G/G^o is etale is therefore equivalent to prove $(G/G^o)^o = e$; if H is the inverse image of $(G/G^o)^o$ in G, then H is an extension of two connected groups; this implies that H is connected (for any field K in $\underline{Mf}_k$

then

$$0 \longrightarrow G^{o}(K) \longrightarrow H(K) \longrightarrow (G/G^{o})(K)$$

is an exact sequence, hence $H(K) = \{0\})$ hence $H \subsetneq G^{o}$ i.e. $H = G^{o}$ and $(G/G^{o})^{o} = e$.

Suppose now k is perfect. Let k_m be the residue field of A_m, and $B = \prod k_m$. Then $\mathrm{Spf}\, B$ is etale and is a subgroup of G (because B is quotient biring of A); put $G^{e} = \mathrm{Spf}\, B$. Then $(G \otimes_k \bar{k})^{e} = G^{e} \otimes_k \bar{k}$ as is readily checked, and G is the product of G^{o} and G^{e}, because this becomes true by going to $\bar{k}$.

An affine group G is said to be infinitesimal if it is finite and local, equivalently, if G is algebraic and $G(\bar{k}) = e$. By the preceding proposition, we see that a finite group is an extension of an etale group by an infinitesimal group and that this extension splits if k is perfect.

Definition. A (not-necessarily commutative) connected formal group $G = \mathrm{Spf}\, A$ is said to be of finite type if A is noetherian; the dimension of G is by definition the Krull dimension of A.

Let m be the maximal ideal of A; it is well known that A is noetherian if and only if $[m/m^2 : k] < +\infty$, and that $\dim G \leq [m/m^2 : k]$.

Lemma ($p \neq 0$). A connected formal group G is of finite type if and only if $\mathrm{Ker}\, F_G$ is finite. If G is of finite type, then $\mathrm{Ker}\, F_G^n$ is finite for all n.

If $\mathrm{Ker}\, F_G$ is finite, then $[A/m^{\{p\}} : k] \leq \infty$, hence $[m/m^2 : k] < +\infty$. Conversely, if m/m^2 is generated by the classes of $X_1, \ldots, X_n$, then A is a quotient of $k[[X_1, \ldots, X_n]]$, and $A/m^{\{p^n\}}$ is a quotient of the finite k-ring $k[[X_1, \ldots, X_n]]/(X_1, \ldots, X_n)^{\{p^n\}}$.

It follows that if $p \neq 0$ *a connected formal group of finite type is an inductive limit of finite groups* ($G = \varinjlim \operatorname{Ker} F_G^n$).

If G is an algebraic group-scheme, then the "connected completion" $\hat{G}^o$ of G is of finite type:

$$\hat{G}^o = \operatorname{Spf} \hat{O}_{G,e} \left[= \varinjlim \operatorname{Ker} F_G^n \text{ if } p \neq 0 \right].$$

8. *Multiplicative affine groups.*

Lemma. Let G *be a* k-*group-functor. Then the following conditions are equivalent:*

(i) G *is the Cartier dual of a constant group.*

(ii) G *is an affine* k-*group and the* k-*ring* $O(G)$ *is generated by the characters of* G (i.e. *homomorphisms from* G *to* μ_k).

If $G = D(\Gamma_k)$, then $G(R) = \underline{Gr}_R(\Gamma_R, \mu_R) = \underline{G}(\Gamma, R^*) = \underline{M}_k(k[\Gamma], R)$, hence $G = \operatorname{Sp} k[\Gamma]$, where $k[\Gamma]$ is the algebra of the group Γ (note that $\Delta\gamma = \gamma \otimes \gamma$, $\varepsilon\gamma = 1$, $\sigma\gamma = \gamma^{-1}$, $\gamma \in \Gamma$), and each $\gamma \in \Gamma \subset k[\Gamma] = O(G)$ is a character of G.

Conversely, if G is affine and $O(G)$ generated by characters, let Γ be the group of all characters of G; then the canonical map $k[\Gamma] \longrightarrow O(G)$ is surjective. But it is always injective (Dedekind's lemma on linear independence of characters), hence $k[\Gamma] \simeq O(G)$.

Such a group is called diagonalizable.

Theorem. Let G *be a* k-*group. Then the following conditions are equivalent:*

(i) $G \otimes_k k_s$ *is diagonalizable.*

(ii) $G \otimes_k K$ *is diagonalizable for a field* $K \in \underline{M}_k$.

(iii) G *is the Cartier dual of an etale* k-*group.*

(iv) $\hat{D}(G)$ *is an etale* k-*formal group.*

(v) $\underline{Gr}_k(G, \alpha_k) = 0.$

(vi) (*If* $p \neq 0$), $V_G : G^{(p)} \longrightarrow G$ *is an epimorphism.*

(vii) (If $p \neq 0$), $V_G : G^{(p)} \longrightarrow G$ *is an isomorphism.*

The implications (i)$\Longleftrightarrow$(iv)$\Longleftrightarrow$(vii)$\Longleftrightarrow$(vi) are clear.

Proof of (v)$\Longleftrightarrow$(iv). We know that $\underline{Gr}_k(G, \alpha_k)$ is the set of primitive elements of $O(G)$; let $A = O(G)$ and let A' be the ring of $\hat{D}(G)$ (i.e. the topological dual of the coring A). By duality, a primitive element of A corresponds to an algebra morphism

$$A' \longrightarrow k[t]/t^2$$

compatible with the augmentations of A' and $k[t]/t^2$. All primitive elements are zero if and only if A'^{o} has no quotients isomorphic to $k[t]/t^2$, which means that $A'^{o} = k$, i.e. $\hat{D}(G)^{o} = e$, i.e. $\hat{D}(G)$ etale.

End of the proof. If k' is an extension of k, then condition (v) for G is equivalent to condition (v) for $G \otimes k'$. This implies the equivalence of all conditions except (iii). It is clear that (iii)$\Longrightarrow$(i) (definition); conversely, if $\hat{D}(G)$ is etale, then let E be the etale k-group such that $\hat{E} = \hat{D}(G)$; we claim that $D(E) \simeq G$. This is easy if $k = k_s$, because E is constant; the general case is proved by going to k_s (see D.G, IV, 1.3.2).

Such a group is called _multiplicative_; the multiplicative groups correspond by duality to etale formal groups; they form a _thick subcategory_ (= stable by subgroups, quotients, extensions) stable for $\varprojlim$, of $\underline{AC}_k$, called $\underline{ACm}_k$, and anti-equivalent to the category of Galois-modules: to $G \in \underline{ACm}_k$ corresponds the Galois-module $X(G) = \hat{D}(G \otimes_k k_s)(k_s) = \underline{Gr}_{k_s}(G \otimes_k k_s, \mu_{k_s})$.

Remark. If E is an etale k-group, then $D(E)$ is multiplicative and $\hat{D}(D(E)) = \hat{E}$; in fact, one already has $D(D(E)) = E$. [D.G., _loc. cit._] It implies that the anti-equivalence between multiplicative groups and etale groups can also be given (without speaking about formal-groups at all) by $E \longrightarrow D(E)$, $G \longrightarrow D(G)$.

9. _Unipotent affine groups. Decomposition of affine groups._

Theorem. Let G _be an affine_ k-_group. The following conditions are equivalent._

(i) $\hat{D}(G)$ _is a connected formal group._

(ii) _Any multiplicative subgroup of_ G _is zero._

(iii) _For any subgroup_ H _of_ G, $H \neq 0$, _we have_ $\underline{Gr}_k(H, \alpha_k) \neq 0$.

(iv) _Any algebraic quotient of_ G _is an extension of subgroups of_ α_k.

(v) (_If_ $p \neq 0$), $\bigcap \operatorname{Im} V_G^n = e$.

The equivalence of (i) and (ii) is clear (the formal group H is connected, iff $\pi_0(H) = e$, i.e. iff it has no etale quotients). The equivalence of (ii) and (iii) follows from the theorem of n°8. The equivalence of (iii) and (iv) is clear because algebraic groups are artinian. Suppose $p \neq 0$. If G satisfies (iv), then for any algebraic quotient H of G, one has $V_H^n = 0$ for large n (recall that $V_{\alpha_k} = 0$). It follows that $\bigcap \operatorname{Im} V_G^n$ has no algebraic quotients, hence is e. Conversely, if (v) is true for G, G cannot contain a non-zero multiplicative subgroup H, for $V_H^n : H^{(p^n)} \longrightarrow H$ is an epimorphism for all n.

Such a group is called <u>unipotent</u>. The unipotent groups correspond by duality to connected formal groups. They form a thick subcategory, stable for $\varprojlim$, of $\underline{AC}_k$, called $\underline{ACu}_k$.

By duality, the theorem of n°7 gives:

<u>Theorem</u>. <u>An affine group is in a unique way an extension of a unipotent group by a multiplicative group. This extension splits if</u> k <u>is perfect</u>.

In particular, if k is <u>perfect</u>, any finite group is uniquely the product of four subgroups which are respectively etale multiplicative, etale unipotent, infinitesimal multiplicative and infinitesimal unipotent. Therefore the category $\underline{F}_k$ of finite (commutative) k-groups splits as a product of four subcategories, called $\underline{Fem}_k$, $\underline{Feu}_k$, $\underline{Fim}_k$, $\underline{Fiu}_k$. The categories $\underline{Feu}_k$ and $\underline{Fim}_k$ are dual to each other, the categories $\underline{Fem}_k$ and $\underline{Fiu}_k$ are autodual.

<u>Proposition</u> 1) <u>Let</u> $p = 0$. <u>Then</u> $\underline{F}_k = \underline{Fem}_k$: <u>any finite</u> (<u>commutative</u>) k-<u>group is etale and multiplicative</u>.

2) <u>Let</u> $p \neq 0$ <u>and</u> k <u>be algebraically closed</u>. <u>Any</u> (<u>commutative</u>) <u>finite</u> k-<u>group is an extension of copies of</u> ${}_p\alpha_k$, ${}_p\mu_k$ <u>and</u> $(\mathbb{Z}/r\mathbb{Z})_k$, r <u>prime</u>.

<u>Proof of</u> 1). By duality, it suffice to prove that any finite unipotent group is zero. Such a group is a product of an etale unipotent group and an infinitesimal unipotent group; by the first theorem, these two groups are extensions respectively of etale subgroups of α_k and infinitesimal subgroups of α_k. Any etale subgroup of α_k must be zero, because $\alpha_k(\bar{k}) = \bar{k}$ has no finite subgroups; an infinitesimal subgroup of α_k is of the form $\mathrm{Sp}\, k[T]/T^n$ where n must be such that $\Delta T^n \subseteq (T^n) \otimes k[T] + k[T] \otimes (T^n)$, this means $(T + T')^n = \alpha T^n + \beta T'^n$ and implies $n = 1$.

<u>Proof of</u> 2). Let $G \in \underline{F}_k$. If G is etale, then $G = \Gamma_k$ where Γ is a finite group; but Γ is an extension of groups $\mathbb{Z}/r\mathbb{Z}$, r prime, and G is an extension

of $(\mathbb{Z}/r\mathbb{Z})_k$. If G is infinitesimal and multiplicative, then $G = D(\Gamma_k)$, where Γ is finite and $\underline{Gr}(\Gamma, \bar{k}^*) = 0$; this implies Γ is p-torsion, and G is an extension of copies of $D((\mathbb{Z}/p\mathbb{Z})_k) = {}_p\mu_k$. If G is infinitesimal and unipotent, then G is an extension of infinitesimal subgroups of α_k. These are the ${}_{p^r}\alpha_k$, because $(T+T')^n = \alpha T^n + \beta T'^n$ implies $n = p^r$; but ${}_{p^r}\alpha_k$ is a p-fold extension of ${}_p\alpha_k$ (remark that ${}_{p^r}\alpha_k / {}_p\alpha_k = {}_{p^{r-1}}\alpha_k$).

Corollary. *If m is a prime, and G a finite (commutative) k-group, then $m^\alpha \cdot id_G = 0$ for large α if and only if $rk(G)$ is a power of m.*

It follows from the multiplicativity of the rank, the fact that $rk(G \otimes_k \bar{k}) = rk(G)$ and the obvious formulas:

$$rk((\mathbb{Z}/r\mathbb{Z})_k) = r, \quad rk({}_p\alpha_k) = rk({}_p\mu_k) = p.$$

In particular, if $p^\alpha \cdot id_G = 0$, then $rk(G) = p^{\text{length}\,(G \otimes_k \bar{k})}$, where length (G) is the length of a Jordan-Holder series of G.

10. Smooth formal-groups.

A (not-necessarily commutative) connected formal group $G = \operatorname{Spf} A$ is said to be smooth if A is a power-series algebra $k[[X_1,\ldots,X_n]]$. In that case, the coproduct $\Delta : A \longrightarrow A \hat{\otimes} A$ is given by a set of formal power series.

$$\Phi(X,Y) = (\Phi_i(X_1,\ldots,X_n, Y_1,\ldots,Y_n)),\ i = 1,\ldots,n$$

and the axioms (Ass) and (Un) give

$$(\text{Ass})\ \Phi(X,\Phi(Y,Z)) = \Phi(\Phi(X,Y),Z)$$

$$(\text{Un})\ \Phi(0,Y) = \Phi(X,0) = 0$$

It is easily proved, using the implicit function theorem, that the existence of an antipodism is a consequence of (Ass) and (Un). The axiom (Com) can be written.

(Com) $$\Phi(X,Y) = \Phi(Y,X).$$

Such a set $\{\Phi_i\}$ is a formal-group-law in the sense of Dieudonné.

Theorem. *Let $G = \operatorname{Spf} A$ be a (not-necessarily commutative) connected formal group of finite type.*

1) *If $p = 0$, then G is smooth.*

2) *If $p \neq 0$, the following conditions are equivalent:*

a) *G is smooth,*

b) *$A \otimes_k k^{p^{-1}}$ is reduced.*

c) *$F_G : G \longrightarrow G^{(p)}$ is an epimorphism.*

Remark first that in 2) we have a) $\Longrightarrow$ b); moreover c) is equivalent to $F_A : A^{(p)} \longrightarrow A$ being injective, or to $A^{(p)} \simeq A \otimes_k k^{p^{-1}}$ being reduced. We then have to prove that if, either $p = 0$, or $p \neq 0$ and $A \otimes_k k^{p^{-1}}$ is reduced, then $A \cong k[[X_1, \dots, X_n]]$.

Let first m be $\operatorname{Ker}(\varepsilon : A \longrightarrow k)$ and $\delta : m/m^2 \longrightarrow k$ be a linear form. We claim that *there exists a continuous k-derivation D of A such that for $a \in m$, one has* $\varepsilon D(a) = \delta(a \bmod m^2)$. Define first $\bar{\delta}(a) = \delta((a - \varepsilon a) \bmod m^2)$; then $\bar{\delta}(ab) = \varepsilon(a)\,\bar{\delta}(b) + \varepsilon(b)\,\bar{\delta}(a)$; put $D = (1 \otimes \bar{\delta}) \circ \Delta$: if $\Delta a = \sum a_i \otimes b_i$, then $Da = \sum a_i \bar{\delta} b_i$. One has $\varepsilon Da = \sum \varepsilon(a_i)\,\bar{\delta}(b_i) = \bar{\delta}(\sum \varepsilon(a_i) b_i) = \bar{\delta} a$; it remains to show that D is a derivation:

$$D(ab) = (1 \otimes \bar{\delta})\Delta(ab) = (1 \otimes \bar{\delta})(\Delta a \Delta b) = (1 \otimes \varepsilon)\Delta a.(1 \otimes \bar{\delta})\Delta b + (1 \otimes \varepsilon)\Delta b.\,(1 \otimes \bar{\delta})\Delta a = aDb + bDa.$$

Let now ξ_i be elements of m such that their classes modulo m^2 form a basis of m/m^2. The canonical map

$$f : k[[X_1,\ldots,X_n]] \longrightarrow A, \quad f(X_i) = \xi_i$$

is surjective. Suppose it is not injective. Let $\Phi \in \operatorname{Ker} f$, $\Phi \neq 0$, <u>with minimal valuation</u>; certainly $v(\Phi) > 0$ (because $\Phi(0) = \varepsilon f(\Phi) = 0$). By the above remark, there exists continuous derivations D_i of A with $D_i(\xi_j) \equiv \delta_{ij} \bmod m$.

Clearly $0 = D_i f(\Phi) = \sum f(\frac{\partial\Phi}{\partial X_j}) D_i(\xi_j)$. But the matrix $(D_i(\xi_j))$ is congruent mod m to the identity matrix, hence is invertible. It follows that $\frac{\partial\Phi}{\partial X_j} = 0$.

If $p = 0$, then Φ must be 0, and f is injective. If $p \neq 0$, then there exists $\psi \in k^{1/p}[[X_1,\ldots,X_n]]$ with $\Phi = \psi^p$; extend f to $f' : k^{1/p}[[X_1,\ldots,X_n]] \longrightarrow A \otimes_k k^{1/p}$; then $f'(\psi)^p = f(\Phi) = 0$. Because $A \otimes_k k^{1/p}$ is reduced, this implies $f'(\psi) = 0$. But Φ was supposed of minimal valuation, hence $\psi = 0$ (if not, decompose ψ as a sum $\sum \lambda_i \psi_i$, $\lambda_i \in k^{1/p}$, $\psi_i \in \operatorname{Ker} f$, $\psi_i \neq 0$, and note that $v(\psi) \geqslant \inf v(\psi_i)$) and $\Phi = 0$.

q.e.d.

The preceding theorem can be strengthened:

1) (Cartier). If $p = 0$, and $G = Sp^*C$ is a connected (not-necessarily commutative) formal-group, then C is the universal enveloping algebra of the Lie algebra $\mathfrak{g}$ of G. This implies that the category of all connected formal-groups is equivalent to the category of all Lie algebras over k. By the Poincaré-Birkhoff-Witt theorem, this also implies that, if $\mathfrak{g}$ is finite dimensional, then G is smooth. Moreover, if G is commutative, then $\mathfrak{g}$ is abelian, hence $G \simeq (\hat{\alpha}^0)^{(I)}$; by duality, any <u>unipotent</u> (<u>commutative</u>) k-<u>group is a power of the additive group</u>.

2) (Dieudonné-Cartier-Gabriel). If $p \neq 0$, k is perfect, G is any (not-necessarily commutative) connected formal group of finite type, H a subgroup, and $G/H = \mathrm{Spf}\, A$ (the quotient which has not been defined in these lectures), then A is of the form $k[[X_1,\ldots,X_n]]\,[Y_1,\ldots,Y_d]/(Y_1^{p^{r_1}},\ldots, Y_d^{p^{r_d}})$. This applies for instance to $A = \hat{O}_{G,e}$, G an algebraic k-group.

Corollary. Suppose $p \neq 0$, and let G be a connected formal group of finite type.

1) If k is perfect, there exists a unique exact sequence of connected groups

$$0 \longrightarrow G_{red} \longrightarrow G \longrightarrow G/G_{red} \longrightarrow 0,$$

with G_{red} smooth, and G/G_{red} infinitesimal (= finite).

2) For large r, the group $G/\mathrm{Ker}\, F_G^r = \mathrm{Im}(G \longrightarrow G^{(p^r)})$ is smooth.

Proof 1) The uniqueness is clear, because any homomorphism from a smooth group to an infinitesimal group is zero (look at the algebras). Let $G = \mathrm{Spf}\, A$, and $G_{red} = \mathrm{Spf}\, A_{red}$, where $A_{red} = A/n$ is the quotient of A by its nilideal. Because $A_{red} \hat{\otimes}_k A_{red}$ is reduced (see the appendix, n° 12),

$$\Delta n \subseteq A \hat{\otimes} n + n \hat{\otimes} A$$

and G_{red} is a subgroup of G, smooth by the theorem. Moreover $G/G_{red} = \mathrm{Spf}\, B$, where $B = \{x \in A, \Delta x - x \otimes 1 \in A \otimes n\}$. If $x \in B$, $\varepsilon(x) = 0$, then $x = \varepsilon \otimes 1\, (\Delta x - x \otimes 1) \in n$. It implies $B \subseteq k + n$, and B is artinian, hence finite.

2) It is clear that $H = G/F^n_G$ is smooth if and only if $H \otimes_k \bar{k}$ is. Replacing k by $\bar{k}$, we can suppose k perfect and apply 1). There exist an i with $F^i(G/G_{red}) = 0$; but $F^i(G_{red}) = G_{red}^{(p^i)}$ because G_{red} is smooth. Hence $F^iG = F^i(G_{red}) = G_{red}^{(p^i)}$ and F^iG is smooth.

<u>Corollary</u>. <u>Let</u> G <u>be a connected formal group of finite type, and</u> $n = \dim G$. <u>Then</u> $\mathrm{rk}(\mathrm{Coker}\, F^i_G)$ <u>is bounded and</u>

$$\mathrm{rk}(\mathrm{Ker}\, F^i_G) = p^{ni}.\ \mathrm{rk}(\mathrm{Coker}\, F^i_G).$$

If G is smooth, then F_G is an epimorphism, and $\mathrm{Ker}\, F^i_G \simeq \mathrm{Spf}\, k[[X_1,\ldots,X_n]]/(X_1,\ldots,X_n)^{\{p^r\}}$, hence $\mathrm{rk}(\mathrm{Ker}\, F^i_G) = p^{ni}$. In the general case, let r be such that $H = F^rG$ is smooth, let $K = \mathrm{Ker}\, F^r_G$; we have exact sequences:

$$0 \longrightarrow \mathrm{Ker}\, F^i_K \longrightarrow \mathrm{Ker}\, F^i_G \longrightarrow \mathrm{Ker}\, F^i_H \longrightarrow \mathrm{Coker}\, F^i_K \longrightarrow \mathrm{Coker}\, F^i_G \longrightarrow 0,$$

$$0 \longrightarrow \mathrm{Ker}\, F^i_K \longrightarrow K \longrightarrow K^{(p^i)} \longrightarrow \mathrm{Coker}\, F^i_K \longrightarrow 0$$

The second sequence gives $\mathrm{rk}(\mathrm{Coker}\, F^i_K) = \mathrm{rk}(\mathrm{Ker}\, F^i_K) \leqslant \mathrm{rk}(K) < \infty$, the first one gives the claimed formula.

<u>Corollary</u> 1) <u>Let</u> $0 \longrightarrow G' \longrightarrow G \longrightarrow G'' \longrightarrow 0$ <u>be an exact sequence of connected formal-groups</u>. <u>Then</u> $\dim(G) = \dim(G') + \dim(G'')$.

2) <u>If</u> $f: G' \longrightarrow G$ <u>is a homomorphism of connected formal group, with</u> G <u>smooth, and</u> $\dim G = \dim G'$, <u>then</u> f <u>is an epimorphism if and only if</u> $\mathrm{Ker}\, f$ <u>is finite</u>.

1) follows from the snake diagram and the preceding corollary.

2) We have the equivalence (Ker f finite) $\Longleftrightarrow$ (dim(Ker f) = 0) $\Longleftrightarrow$ (dim f(G') = dim G') $\Longleftrightarrow$ (dim f(G') = dim G). But dim f(G') = dim G gives

$$\operatorname{rk} \operatorname{Ker} F^i_{f(G')} \geqslant p^{i \dim G} = \operatorname{rk}(\operatorname{Ker} F^i_G),$$

hence $\operatorname{Ker} F^i_{f(G')} = \operatorname{Ker} F^i_G$, and $G = \bigcup \operatorname{Ker} F^i_G = \bigcup \operatorname{Ker} F^i_{f(G')} = f(G')$.

11. p-divisible formal groups.

Suppose $p \neq 0$.

Definition. A (commutative) formal group G is called p-divisible (or a Barsotti-Tate group) if it satisfies the three following properties:

1) $p.\mathrm{id}_G : G \longrightarrow G$ is an epimorphism,

2) G is a p-torsion group: $G = \bigcup_j \operatorname{Ker}(p^j.\mathrm{id}_G)$,

3) $\operatorname{Ker}(p.\mathrm{id}_G)$ is finite.

We know that $\operatorname{rk}(\operatorname{Ker} p\, \mathrm{id}_G) = p^h$, $h \in \mathbb{N}$. This h is called the height ht(G) of G. Using 1), this gives

$$\operatorname{rk}(\operatorname{Ker} p^j\, \mathrm{id}_G) = p^{j.\mathrm{ht}(G)}.$$

The multiplicativity of the rank gives the exactness of the sequences

$$0 \longrightarrow \operatorname{Ker} p^j \xrightarrow{\text{inclusion}} \operatorname{Ker} p^{j+k} \xrightarrow{p^j} \operatorname{Ker} p^k \longrightarrow 0$$

Conversely, if we have a diagram

$$G_1 \xrightarrow{i_1} G_2 \xrightarrow{i_2} G_3 \longrightarrow \dots .$$

where the G_i are finite k-groups with the following properties.

a) $rk(G_j) = p^{hj}$, h a fixed integer,

b) the sequences $0 \longrightarrow G_j \xrightarrow{i_j} G_{j+1} \xrightarrow{p^j} G_{j+1}$ are exact,

then $\varinjlim (G_n, i_n)$ a p-divisible formal group, of height h, and $Ker(p^n \, id_G : G \longrightarrow G) \simeq G_n$.

This gives an <u>alternative definition of p-divisible groups</u>.

The (Serre) dual of a p-divisible group G is the p-divisible group G' defined as follows:

Let $G_j = Ker(p^j \, id_G)$, and let $p_j : G_{j+1} \longrightarrow G_j$ be induced by $p \, id_G$. Put $G'_j = D(G_j)$, and $i'_j = D(p_j) : G'_j \longrightarrow G'_{j+1}$, then $G' = \varinjlim (G'_j, i'_j)$ is a p-divisible formal group, with $ht(G') = ht(G)$; it is clear that $p'_j = D(i_j)$, so that $(G')'$ can be identified with G.

<u>Examples</u> 1) The constant formal group $(\mathbb{Q}_p/\mathbb{Z}_p)_k$ is a p-divisible group of height 1; conversely, any constant p-divisible group of height h is isomorphic to $(\mathbb{Q}_p/\mathbb{Z}_p)^h_k$.

2) Let A be a (commutative) algebraic k-group, such that $p id_G : A \longrightarrow A$ is an epimorphism. Then, it can be shown that $Ker(p.id_A)$ is finite; define

$$A(p) = \bigcup_j Ker(p^j \, id_A).$$

Then $A(p)$ is a p-divisible group, containing $\hat{A}^o = \bigcup_j Ker(F^j_G)$. For instance, for $A = \mu_k$, one finds $A(p) = \bigcup_j {}_{p^j}\mu_k = (\mathbb{Q}_p/\mathbb{Z}_p)'_k$.

If A is an abelian variety of dimension g, one knows that $p\, id_A$ is an epimorphism, with $rk(Ker\ p\, id_G) = p^{2g}$. It follows that $A(p)$ is a p-divisible group of height $2g$ (see Chapter V).

Proposition. Let G be a k-formal group. Then G is p-divisible if and only if the following conditions are satisfied.

1) $\pi_0(G)(\bar{k}) \simeq (\mathbb{Q}_p/\mathbb{Z}_p)^r$, r finite.

2) G^o is of finite type, smooth, and $Ker(V: G^{o(p)} \longrightarrow \overset{o}{G})$ is finite.

If G is p-divisible, then G^o and $\pi_0(G)$ are, and conversely (replace k by $\bar{k}$, then G is the product of G^o and $\pi_0(G)$). We already know that the etale group E is p-divisible iff $E(\bar{k}) \simeq (\mathbb{Q}_p/\mathbb{Z}_p)^r$. We therefore can suppose G connected.

Suppose G is p-divisible, then $Ker\ F_G \subset Ker(V_G F_G) = Ker(p\, id_G)$ is finite, hence G is of finite type: on the other hand $G^{(p)}$ also is p-divisible, hence $Ker\ V_G \subseteq Ker(F_G V_G) = ker\ (p\, id_{G^{(p)}})$ is finite, and F_G is an epimorphism, because $p\, id_G(V) = F_G V_G$ is.

Conversely, if G is smooth, and $Ker\ V_G$ finite, F_G and V_G are epimorphisms (n° 9), hence also $p\, id_G = V_G F_G$; this implies also an exact sequence

$$0 \longrightarrow Ker\ (F_G) \longrightarrow Ker\ (p\, id_G) \longrightarrow Ker\ (V_G) \longrightarrow 0$$

and $Ker\ (p\, id_G)$ also is finite. Finally $\bigcup Ker\ (p^j\, id_G) \supseteq \bigcup Ker\ (F_G^j) = G$.

Example. If A is an algebraic unipotent k-group, then $\hat{A}^o$ is never p-divisible, unless A is finite.

Remark. The above exact sequence gives for any p-divisible group G the formula height $(G) = \dim (G) + \dim (G')$.

Proposition. Let G *be a connected, finite type, smooth formal group. There exist two subgroups* $H, K \subseteq G$ *with* H *p-divisible,* $p^n K = 0$ *for large* n, $H \cap K$ *finite*, and $G = H + K$.

Let $p^n G = \operatorname{Im}(p^n \, id_G : G \longrightarrow G)$; the subgroups $p^n G$ of G are smooth (quotients of G) and form a decreasing sequence. There exist an n such that $p^n G \cap \operatorname{Ker} F_G = p^{2n} G \cap \operatorname{Ker} F_G$ ($\operatorname{Ker} F_G$ is finite, hence artinian). This implies $p^n G = p^{2n} G$, because $p^n G / p^{2n} G$ is connected, smooth, with monomorphic Frobenius (or dimension zero). Put $H = p^n G$, $K = \operatorname{Ker}(p^n \, id_G)$. Then $G = H + K$, $p \, id_H$ is epimorphic, and $p^n K = 0$. Therefore $\operatorname{Ker}(p \, id_H)$ is finite, hence H is p-divisible and $H \cap K \subseteq \operatorname{Ker}(p^n \, id_H)$ is finite.

12. Appendix.

Theorem. Let k *be perfect field with characteristic* $p \neq 0$, A *and* B *two complete noetherian k-rings with residue field* k. *If* A *and* B *are reduced, so is* $A \hat{\otimes}_k B$.

1) Let α be a positive integer. We say that a k-ring R has property (N_α) if R is local artinian with residue field k, and if $x \in R$, $x^p = 0$ implies $x \in \mathfrak{m}_R^\alpha$ ($\mathfrak{m}_R$ = maximal ideal of R).

Lemma 1. If R *and* S *have property* (N_α), *so has* $R \otimes_k S$.

Let x_i be a basis of the k-vector space R such that the $x_i \in \mathfrak{m}_R^r$ are a basis of $\mathfrak{m}_R^r$ for all r. Let $z \in R \otimes S$, with $z^p = 0$; we can write $z = \sum x_i \otimes y_i$, hence $\sum x_i^p \otimes y_i^p = 0$. This implies the existence of elements $\lambda_{i,j} \in k$ and $s_j \in S$ with

$$\sum_i \lambda_{i,j} \, x_i^p = 0, \; y_i^p = \sum_j \lambda_{i,j} s_j$$

Because k is perfect, each $\lambda_{i,j}$ can be written as $\mu_{i,j}^p$ and we have $(\sum \mu_{i,j}\, x_i)^p = 0$, hence $\sum \mu_{i,j}\, x_i \in m_R^\alpha$, hence $\mu_{i,j} = 0$ for $x_i \notin m_R^\alpha$. If $x_i \notin m_R^\alpha$, then $\mu_{ij} = 0$ for all j, hence $y_i = 0$, hence $y_i \in m_S^\alpha$; in any case $x_i \otimes y_i \in m_R^\alpha \otimes S + R \otimes m_S^\alpha \subseteq (m_{R\otimes S})^\alpha$, and $z \in m_{R\otimes S}^\alpha$.

2) Let A be a local complete noetherian k-ring with residue field k. Put $A_r = A/m_A^r$, and let $\alpha_A(r)$ be the greatest α such that A_r has property (N_α): $\alpha_A(r)$ is the greatest integer such that

$$x \in A,\ x^p \in m_A^r \Longrightarrow x \in m_A^{\alpha_A(r)} .$$

Then $\alpha_A(1) \leqslant \alpha_A(2) \leqslant \dots \leqslant \alpha_A(r) \leqslant \dots$.

<u>Lemma 2</u>. A <u>is reduced iff</u> $\lim_r \alpha_A(r) = +\infty$.

If $x \in A$ with $x^p = 0$, $x \in m_A^N$, $x \notin m_A^{N+1}$, then $\alpha_A(r) \leqslant N$ for all r.

Conversely, suppose A is reduced, let $V_i = \{x \in A,\ x^p \in m_A^i\}$. Then (V_i) is a decreasing sequence of ideals of A, and $\cap V_i = 0$. By definition, $\alpha(r)$ is the greatest integer with $V_r \subseteq m_A^{\alpha(r)}$, and $\cap V_i = 0$ implies $\lim_r \alpha(r) = \infty$ (Chevalley's theorem, <u>see</u> Zariski-Samuel, Chapter VIII, § 5).

3) Let now A and B be as in the theorem and put $C = A \hat{\otimes} B$, then lemma 1 gives

$$\alpha_C(r) \geqslant \inf(\alpha_A(r), \alpha_B(r)),$$

and we conclude by Lemma 2.

CHAPTER III

WITT GROUPS AND DIEUDONNE MODULES

Let p be a fixed prime number.

1. The Artin-Hasse exponential series.

Let k be a ring. We denote by Λ_k the affine k-group which associates with $R \in \underline{M}_k$ the multiplicative group $1 + tR[[t]]$ of formal power-series in R with constant term 1 (as a k-functor, Λ_k is obviously isomorphic to $\underline{O}_k^{\mathbb{N}}$). For $n \geq 1$, let $\Lambda_k^{(n)}$ be the closed subgroup such that

$$\Lambda_k^{(n)}(R) = 1 + t^n R[[t]] = \{1 + a_n t^n + \dots\} \quad ;$$

one has obvious exact sequences

$$0 \longrightarrow \Lambda_k^{(n+1)} \longrightarrow \Lambda_k^{(n)} \longrightarrow \underline{\alpha}_k \longrightarrow 0$$

where the first morphism is the inclusion, the second one being $(1+a_n t^n + \dots) \longrightarrow a_n$. The k-group Λ_k hence appears as the inverse limit of the $\Lambda_k/\Lambda_k^{(n+1)}$, each $\Lambda_k/\Lambda_k^{(n+1)}$ being an n-fold extension of the additive group. (If k is a field, then Λ_k is a unipotent group).

Let $F = 1 - t + \dots$ be a fixed element of $\Lambda(k) = 1 + tk[[t]]$. Then we have an isomorphism of k-schemes (where $\mathbb{N}_+ = \{1,2,\dots\}$).

$$\varphi : \underline{O}_k^{\mathbb{N}_+} \longrightarrow \Lambda_k$$

by $\varphi((a_n)) = \prod F(a_n t^n)$.

If $k = \mathbb{Q}$, then take $F(t) = \exp(-t)$; one has $F(at)F(bt) = F((a+b)t)$, so that φ is an <u>isomorphism</u> of k-<u>groups</u> from $\alpha_k^{N_+}$ to Λ_k. If k is a field with characteristic p, it is not possible to find $F \in 1 + tk[[t]]$ with

$$F(t) = 1 - t + \dots, F(at)F(bt) = F(ct);$$

we find first $F(T) = 1 - t + \dots + (-t)^{p-1}/(p-1)! + \dots$ and for the coefficient of T^p we find $0 = 1$ and the computation fails. But remark that for any F one certainly has a formula

$$F(at)F(bt) = \prod_{i>0} F(\lambda_i(a,b)t^i); \tag{1}$$

where $\lambda_i(X,Y) \in k[X,Y]$.

The idea is to find an F such that most of the λ_i vanish. Actually we shall find F with $\lambda_i = 0$ if i is not a power of p.

A classical formula asserts

$$\exp(-t) = \prod_n (1 - t^n)^{\mu(n)/n} \tag{2}$$

where μ is the Moebius function. Recall first that $\mu(n) = 0$ if n is divisible by the square of a prime , $\mu(p_1 \dots p_k) = (-1)^k$ if $p_1, \dots, p_k$ are distinct primes and $\mu(1) = 1$; for $n > 1$, one has

$$\sum_{d|n} \mu(d) = 0.$$

It follows that

$$-t = \sum_{n \geqslant 1} -\frac{1}{n} t^n \sum_{d|n} \mu(d) = \sum_{d \geqslant 1} \frac{\mu(d)}{d} \sum_m -\frac{1}{m} t^{dm}$$

$$= \sum_{d \geqslant 1} \frac{\mu(d)}{d} \log(1-t^d),$$

which gives (2). Let

(3) $$F(t) = \prod_{(n,p)=1} (1-t^n)^{\mu(n)/n} = 1-t+\dots ;$$

if $\mathbb{Z}_{(p)} = \{a/b \in \mathbb{Q},\ (p,b) = 1\}$, then

(4) $$F(t) \in \Lambda(\mathbb{Z}_{(p)}).$$

If $\mu(n) \neq 0$, then either $(n,p) = 1$, or $n = pn'$, $(n',p) = 1$. It follows from (2) and (3) that $\exp(-t) = F(t)/F(t^p)^{1/p}$, or

$F(t) = \exp(-t)F(t^p)^{1/p} = \exp(-t-t^p/p)F(t^{p^2})^{1/p^2} = \dots$, so that

(5) $$\begin{cases} F(t) = \exp L(t), \text{ with} \\ L(t) = -t - t^p/p - t^{p^2}/p^2 - \dots - t^{p^i}/p^i - \dots \end{cases}$$

The formula (1) for F can be written $L(at) + L(bt) = \sum L(\lambda_i(a,b)t^i)$ where $\lambda_i \in \mathbb{Z}_{(p)}[X,Y]$. Going to $\mathbb{Q}$, it follows immediately that $\lambda_i = 0$ if i is not a power of p, which give a formula

(6) $$F(at)F(bt) = \prod_{i \geq 0} F(\psi_i(a,b)t^{p^i}).$$

The <u>Artin-Hasse</u> exponential is defined as the morphism

$$E: O^{\mathbf{N}}_{\mathbb{Z}_{(p)}} \longrightarrow \Lambda_{\mathbb{Z}_{(p)}}$$

such that

(7) $$E((a_0,\dots,),t) = \prod_{n \geq 0} F(a_n t^{p^n}).$$

From (6), it follows easily that there exists formula

$$(8)\qquad E((a_i),t)\cdot E((b_i),t) = E((S_i(a_0,..,a_i,b_0,..,b_i),t)$$

where $S_i \in \mathbb{Z}_{(p)}[X_o,\dots,X_i,\ Y_o,\dots,Y_i]$. Moreover because of (7), any $P\in\Lambda(R)$, $R\in \underline{M}_{\mathbb{Z}_{(p)}}$, can be uniquely written

$$P(t) = \prod_{(n,p)=1} E(\vec{a}_n,t^n),$$

with $\vec{a}_n \in R^{\mathbb{N}}$. From this and (10), it follows

Proposition. The $\mathbb{Z}_{(p)}$-group $\Lambda_{\mathbb{Z}_{(p)}}$ is isomorphic to the $\{n/(n,p)=1\}$-power of the subgroup image of E.

By base-change a similar statement applies to $\Lambda_{\mathbb{F}_p}$; it shows that the Artin-Hasse exponential plays over $\mathbb{F}_p$ a somewhat similar role as the usual exponential over $\mathbb{Q}$.

2. The Witt rings (over $\mathbb{Z}$).

By (5) and (7), we can write

$$(9)\qquad E((a_0,\dots),t) = \exp\Big(-\sum_{n\geq 0} t^{p^n}\Phi_n/p^n\Big),$$

with

$$(10)\qquad \Phi_n(a_0,\dots) = a_0^{p^n} + p\,a_1^{p^{n-1}} + \dots + p^n a_n\,.$$

The formula (8) can also be written

(11) $$\Phi_n(a_0,\dots,a_n) + \Phi_n(b_0,\dots,b_n) = \Phi_n(S_o,\dots,S_n).$$

Lemma. We have $S_n \in \mathbb{Z}[X_o,\dots,X_n]$.

We already know that the coefficients of S_i lie in $\mathbb{Z}_{(p)} \subset \mathbb{Q}$. On the other hand, it is clear from (10) that they lie in $\mathbb{Z}[p^{-1}]$. But $\mathbb{Z}_{(p)} \cap \mathbb{Z}[p^{-1}] = \mathbb{Z}$.

Theorem. There exists a unique commutative group law on $O^{\mathbb{N}}_{\mathbb{Z}}$ with the following equivalent properties:

(i) $E : \underline{O}^{\mathbb{N}}_{\mathbb{Z}} \otimes_{\mathbb{Z}} \mathbb{Z}_{(p)} \longrightarrow \Lambda_{\mathbb{Z}_{(p)}}$ is a homomorphism.

(ii) Each $\Phi_n : \underline{O}^{\mathbb{N}}_{\mathbb{Z}} \longrightarrow \underline{\alpha}_{\mathbb{Z}}$ is a homomorphism.

Each (i), (ii) is equivalent to the fact that (with $\dot{+}$ for the law we are constructing)

(12) $$(a_n) \dot{+} (b_n) = (S_n(a_0,\dots, a_n, b_0,\dots b_n)).$$

Hence the uniqueness; it remains to be shown that the law defined by (12) is a commutative group law with unit element $(0,0,\cdots)$. The associativity, commutativity and unit element axioms can be expressed by polynomials identities, with coefficients in $\mathbb{Z}$, in the coefficients of the S_i. These identities are satisfied after going from $\mathbb{Z}$ to $\mathbb{Z}[p^{-1}]$, because the $\Phi_n \otimes_{\mathbb{Z}} \mathbb{Z}[p^{-1}]$ define an isomorphism $O^{\mathbb{N}}_{\mathbb{Z}[p^{-1}]} \longrightarrow O^{\mathbb{N}}_{\mathbb{Z}[p^{-1}]}$. Because $\mathbb{Z} \subset \mathbb{Z}[p^{-1}]$, we are done. The existence of an inverse element can be proved if $p \neq 2$ by the remark that $\varphi_n(-X_o,-X_1,\dots) = -\varphi_n(X_o,X_1,\dots)$; in the general case, the antipodism over $\mathbb{Z}[p^{-1}]$ is given by polynomials with coefficients in $\mathbb{Z}[p^{-1}]$; but these coefficients are also in $\mathbb{Z}_{(p)}$, hence are in $\mathbb{Z}$.

The $\mathbb{Z}$-scheme $O^{\mathbb{N}}_{\mathbb{Z}}$, together with the above law, is called the $\mathbb{Z}$-group of Witt vectors of infinite length relative to p and denoted by W.

If $w = (a_n) \in W(R) = R^{\mathbb{N}}$, a_n is the nth-component of w and $\Phi_n(w)$ the nth-phantom-component of w. The phantom components define a group isomorphism from $W \otimes_{\mathbb{Z}} \mathbb{Z}[p^{-1}]$ to $\underline{\alpha}^{\mathbb{N}}_{\mathbb{Z}[p^{-1}]}$.

Let $T: W \longrightarrow W$ be the monomorphism defined by

(13) $$T((a_0,\dots,a_n,\dots)) = (0,a_0,a_1,\dots).$$

Then $\Phi_o(Tw) = 0$, $\Phi_n(Tw) = p\,\Phi_{n-1}(w)$, $n \geq 1$; it follows that T is group-homomorphism, called the translation. We define the group W_n of Witt-vectors of length n by the exact sequence of group functors

(14) $$0 \longrightarrow W \xrightarrow{T^n} W \xrightarrow{R_n} W_n \longrightarrow 0$$

(i.e. by $W_n(R) = \operatorname{Coker} T^n(R)$ for each R). By the definition of the group law in W, it is clear that $(a_0,a_1,\dots) = (a_0,\dots,a_{n-1},0,\dots) \dotplus T^n(a_n,a_{n+1},\dots)$, which means that as a scheme, W_n is $\underline{O}^n_k$, the projection morphism $W \longrightarrow W_n$ being $(a_0,\dots) \longrightarrow (a_0,\dots,a_{n-1})$. The group law on W_n is $(a_0,\dots,a_{n-1}) \dotplus (b_0,\dots,b_{n-1}) = (S_o(a_0,b_0),\dots,S_{n-1}(a_0,\dots,a_{n-1}\; b_0,\dots,b_{n-1}))$ in particular $W_1 = \underline{\alpha}$. The snake diagram gives from (14) translation homomorphism $T: W_n \longrightarrow W_{n+1}$, such that $T(a_0,\dots,a_{n-1}) = (0,a_0,\dots,a_{n-1})$, projection homomorphisms $R: W_{n+1} \longrightarrow W_n$ such that $R(a_0,\dots,a_n) = (a_0,\dots,a_{n-1})$ and exact sequences

(15) $$0 \longrightarrow W_m \xrightarrow{T^n} W_{n+m} \xrightarrow{R^m} W_n \longrightarrow 0.$$

Moreover, the projections $W \longrightarrow W_n$ give rise to an isomorphism

$$W \simeq \varprojlim W_n .$$

Let $\tau: \underline{O}_{\mathbb{Z}} \longrightarrow W$ be the morphism $a \longrightarrow (a,0,\dots)$. We have $\varphi_n \tau(a) = a^{p^n}$, $E(\tau(a),t) = F(at)$.

Theorem. There exists a unique ring-structure on the $\mathbb{Z}$-group W such that either the two following conditions is satisfied.

(i) each $\Phi_n : W \longrightarrow \underline{O}_{\mathbb{Z}}$ is a ring-homomorphism.

(ii) $\tau(ab) = \tau(a)\,\tau(b)$, $a,b \in R \in \underline{\underline{M}}_{\mathbb{Z}}$.

We first replace $\mathbb{Z}$ by $P = \mathbb{Z}[p^{-1}]$. Then $(\Phi_n) : W_P \longrightarrow \alpha_P^{\mathbb{N}}$ is an isomorphism, hence the existence and uniqueness of a ring structure on W_P satisfying (i); moreover, because $(\Phi_n(\tau(a)) = (a^{p^n})$, this ring-structure satisfies (ii); conversely, consider a ring structure on the P-group $\alpha_P^{\mathbb{N}}$ such that $(a^{p^n}).(b^{p^n}) = ((ab)^{p^n})$; the multiplication is given by polynomials of the form $(x_n)\cdot(y_n) = (\sum a_{ij}^{(n)} x_i y_j)$, with $\sum a_{ij}^{(n)} a^{p^i} b^{p^j} = (ab)^{p^n}$; this gives $a_{ij}^{(n)} = 0$ except when $i = j = n$, and $(x_n)\cdot(y_n) = (x_n y_n)$. This ends the proof for P.

The multiplication in W_P we just found is given by polynomials $M_n(X_o,\dots,X_n, Y_o,\dots,Y_n) \in P[X_o,\dots,Y_n,\dots]$:

$$(a_o,\dots) * (b_o,\dots) = (M_n(a_o,\dots,b_o,\dots));$$

by definition, $\Phi_i((M_n)) = \Phi_i((X_n))\cdot \Phi_i((Y_n))$, $i = 0,\dots$. An easy lemma (D.G.V, § 1.2) proves that $M_n \in \mathbb{Z}[X_o,\dots,Y_o,\dots,]$; the above formula defines then a $\mathbb{Z}$-morphism $W \times W \longrightarrow W$. The fact that it gives a ring structure satisfying (i) and (ii), with unit element $\tau(1) = (1,0,\dots)$ can be expressed by identities between polynomials with coefficients in $\mathbb{Z}$; these identities are true over P and $\mathbb{Z} \longrightarrow P$ is injective.

The $\mathbb{Z}$-ring W is called the Witt ring, each W_n is a quotient ring of W, the canonical morphisms $R_n : W \longrightarrow W_n$ and $R : W_{n+1} \longrightarrow W_n$ are ring-homomorphism (but not T!).

3. The Witt rings (over k).

From now on, k is a field, with characteristic p. We denote by W_k, W_{nk}, the k-rings $W \otimes_{\mathbb{Z}} k$, $W_{nk} \otimes_{\mathbb{Z}} k$; remark that the phantom-components $W_k \longrightarrow \alpha_k$ are now $(a_n) \longmapsto a_o^{p^n}$ (hence the name).

Because $W_k = W_{\mathbb{F}_p} \otimes_{\mathbb{F}_p} k$, we can identify $W_k^{(p)}$ and W_k and the Frobenius morphism $F: W_k \longrightarrow W_k$ is given by

$$F(a_o,\ldots,a_n,\ldots) = (a_o^p,\ldots,a_n^p,\ldots).$$

It is a ring-homomorphism (because F commutes with products). Similar statements are true for Λ_k and the W_{nk}.

Proposition a). The Verschiebung morphism of Λ_k is $\varphi(t) \longrightarrow \varphi(t^p)$, the Verschiebung morphism of W_k is T, the Verschiebung morphism of W_{nk} is $R.T = T.R$.

b) If $x,y \in W_k(R)$, $R \in \underline{M}_k$, then $V(Fx.y) = x.Vy$.

a) If $\varphi = 1 + \sum c_n t^n \in \Lambda(R)$, then $F\varphi = 1 + \sum c_n^p t^n$, and $(F\varphi)(t^p) = 1 + \sum c_n^p t^{np} = \varphi^p = V(F\varphi)$. But F is an epimorphism, hence $V\psi = \psi(t^p)$, for all ψ.

On the other hand, the definition of E and T shows that

$$(16) \qquad E(Tx,t) = E(x,t^p);$$

but $E(x,t^p) = VE(x,t) = E(Vx,t)$ and E is monomorphism, hence $Vx = Tx$. Projecting this formula on W_{nk}, we find $V_{W_{nk}} = R.T = T.R$.

b) Because $F: W_k \longrightarrow W_k$ is an epimorphism, we can suppose $y = Fz$. Then $V(Fx.y) = V(Fx.Fz) = VF(xz) = pxz = x.pz = x.VFz = x.Vy$.

Corollary. *If* $x, y \in W_k(R)$, *then*

$$E(x.Vy,t) = E(Fx.y,t^p).$$

Corollary. If $x = (a_0,\dots,a_n,\dots) \in W_k(R)$, then $px = (0,a_0^p,\dots,a_n^p,\dots)$.

Corollary. *Suppose* k *is perfect; then* W(k) *is a discrete valuation ring, complete, and* $W(k)/pW(k) = k$.

One has $FW(k) = W(k)$ because k is perfect, hence

$$p^n W(k) = T^n F^n W(k) = T^n W(k) \quad \text{and} \quad W(k) = \varprojlim W(k)/p^n W(k).$$

Moreover $W(k)/pW(k) = W_1(k) = \alpha(k) = k$.

Proposition (Witt). *Let* k *be perfect, and let* A *be complete noetherian local with residue field* k. *Let* $\pi: A \longrightarrow k$ *be the canonical projection. There exist a unique ring-homomorphism*

$$u: W(k) \longrightarrow A$$

compatible with the projections $W(k) \longrightarrow k$ *and* π. *If moreover* A *is a discrete valuation ring with* $p.1_A \neq 0$, *then* A *is a free finite* W(k)-*module of rank* $[A/pA:k]$; *in particular if* $pA = A$, *then* u *is an isomorphism.*

Proof. (After Cartier). Consider the ring-morphisms given by the phantom components $\Phi_n: W_{n+1}(A) \longrightarrow A$. If m is the maximal ideal of A, then $\Phi_n((x_n)) \in m^{n+1}$ if $x_i \in m$; this gives a commutative square

$$\begin{array}{ccc} W_{n+1}(A) & \xrightarrow{\Phi_n} & A \\ {\scriptstyle W_{n+1}(\pi)}\downarrow & & \downarrow{\scriptstyle \text{can.}} \\ W_{n+1}(k) & \xrightarrow{\Phi_n} & A/m^{n+1} \end{array}$$

Let $\sigma : k \longrightarrow k$ be given by $\sigma(\lambda) = \lambda^{1/p}$ and put $u_n = \Phi_n \circ W_{n+1}(\sigma^n)$; then, if $a_0, \dots, a_n \in A$

$$u_n(\pi(a_0^{p^n}), \dots, \pi(a_n^{p^n})) = a_0^{p^n} + pa_1^{p^{n-1}} \dots + p^n a_n \text{ mod. } m^{n+1}.$$

Let

$$u = \varprojlim u_n : W(k) \longrightarrow A.$$

Then u is a ring-morphism and $\pi u(\alpha_0, \dots, \alpha_n) = \alpha_0$. This gives the existence of u. Let $u' : W(k) \longrightarrow A$ be another such homomorphism; then $\tau' = u'\tau : k \longrightarrow A$ is compatible with multiplication and such that $\pi\tau' = \mathrm{Id}$; such a τ' is unique, as is well-known (because $\tau'(\alpha)$ must be in $\bigcap (\pi^{-1}(\alpha^{p^{-n}}))^{p^n}$ which has only one element (Cauchy)); on the other hand, any $x \in W(k)$ can be written

$$x = (\alpha_0, \alpha_1, \dots) = (\alpha_0, 0, 0 \dots) \dotplus (0, \alpha_1, 0 \dots) \dotplus (0, 0, \alpha_2, \dots) + \dots$$

$$= \tau(\alpha_0) + p\tau(\alpha_1^{1/p}) + p^2\tau(\alpha_2^{1/p^2}) + \dots$$

and $u'(x)$ must be $\tau'(\alpha_0) + p\tau'(\alpha_1^{1/p}) + p^2\tau'(\alpha_2^{1/p^2}) + \dots$, hence the unicity of u.

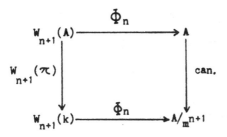

The last statement follows from the fact that if $a_1, \ldots, a_e \in A$ are a basis of A modulo pA, then they generate the $W(k)$-modulo A, (Bourbaki, Alg. Comm. Chap. III, § 2, Prop. 12, Cor. 3). Therefore A is finitely generated as $W(k)$-module, without torsion because $p^n.1_A \neq 0$, hence free of rank $[A/_{pA}:k]$.

4. Duality of finite Witt groups

For $m, n \geq 1$, we put

$$_mW_n = \mathrm{Ker}\,(F^m : W_{nk} \longrightarrow W_{nk}).$$

Between these finite k-groups, we have homomorphisms

$$\begin{array}{ccc} {}_mW_n & \xrightarrow{\;t\;} & {}_mW_{n+1} \\ {\scriptstyle f}\downarrow & & \downarrow{\scriptstyle r} \\ {}_{m-1}W_n & \xrightarrow{\;i\;} & {}_mW_n \, , \end{array}$$

where i is the canonical inclusion, and f,t,r are induced by F,T,R. Clearly, i and t are monomorphisms, f and r are epimorphisms, and for the group $_mW_n$, we have $F = if$, $V = rt$.

For any $R \in \underline{\underline{M}}_k$, let $W'(R)$ be the set of all $(\alpha_0, \alpha_1, \ldots) \in W_k(R)$ such that $a_n = 0$ for large n, and a_n nilpotent for all n. It is easy to check that $W'(R)$ is an ideal in $W_k(R)$ and that $E(w,t)$ is a <u>polynomial</u> for $w \in W'(R)$; in particular $E(w,1)$ is defined for $w \in W'(R)$, and we have a <u>group-homomorphism</u>

$$\widetilde{E} : W' \longrightarrow \mu_k$$

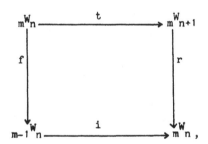

given by $w \longmapsto E(w,1)$. If $x \in W_k(R)$, $y \in W'(R)$, then $xy \in W'(R)$ and $E(xy,1) \in R^*$; moreover, one has

$$E(T^n x.y,1) = E(T^n(x.F^n y),1) = E(x.F^n y,1).$$

The morphism $(x,y) \longmapsto E(xy,1)$ from $W_k \times W'$ to μ_k is bilinear, hence gives a group-homomorphism $W' \longrightarrow D(W_k)$. [This can be shown to be an isomorphism (D.G. V § 4.45) but we shall not need this fact].

Let $\sigma_n : W_{nk} \longrightarrow W_k$ be the section of $R_n : W_k \longrightarrow W_{nk}$ defined by $\sigma_n(\alpha_o,\ldots,\alpha_{n-1}) = (\alpha_o,\ldots,\alpha_{n-1},0,\ldots)$ [σ_n is not a group homomorphism]; it is clear that σ_n sends ${}_mW_n$ in W'.

__Theorem.__ __For__ $x \in {}_mW_n(R)$, $y \in {}_nW_m(R)$, __define__

$$\langle x,y\rangle = E(\sigma_n(x)\,\sigma_m(y),\ 1).$$

__Then__ $\langle x,y\rangle$ __is bilinear, gives an isomorphism__

$${}_mW_n \cong D({}_nW_m)$$

__and satisfies__

$$\langle x,ty\rangle = \langle fx,y\rangle$$

$$\langle x,ry\rangle = \langle ix,y\rangle.$$

Let $x,x' \in {}_mW_n(R)$, $y \in {}_nW_m(R)$; then $\sigma_n(x+x') - \sigma_n(x) - \sigma_n(x')$ is in $\operatorname{Ker} R_n = \operatorname{Im} T^n$, hence

$$\sigma_n(x+x') = \sigma_n(x) + \sigma_n(x') + T^n(u),$$

where $u \in W'(R)$. This implies $\langle x + x', y\rangle = \langle x,y\rangle + \langle x',y\rangle +$ $E(T^n(u)\cdot \sigma_m(y),1)$; but $E(T^n(u)\cdot \sigma_m(y),1) = E(u.F^n \sigma_m(y),1)$, and $F^n\sigma_m(y) = \sigma_m(F^n y) = 0$. This proves the bilinearity of $\langle\, ,\, \rangle$.

On the other hand, $\sigma_n(fx) = F\sigma_n(x)$, $\sigma_{m+1}(ty) = T\sigma_m(y)$, hence $\langle fx,y\rangle = E(F\sigma_n(x)\cdot\sigma_m(y),1) = E(\sigma_n(x)\cdot T\sigma_m(y),1) = \langle x,ty\rangle$; also $\sigma_n(ix) = \sigma_n(x)$, $\sigma_m(ry) = \sigma_{m+1}(y)$, hence $\langle ix,y\rangle = \langle x,ry\rangle$.

It remains to prove that $\langle\, ,\, \rangle$ gives an isomorphism between ${}_mW_n$ and $D({}_nW_m)$; but, because of the exact sequences

$$0 \longrightarrow {}_mW_n \xrightarrow{\;i^q\;} {}_{m+q}W_n \xrightarrow{\;f^m\;} {}_qW_n \longrightarrow 0$$

and

$$0 \longrightarrow {}_nW_m \xrightarrow{\;t^q\;} {}_nW_{m+q} \xrightarrow{\;r^m\;} {}_nW_q \longrightarrow 0$$

and the adjointness of t and f, and r and i, we are reduced by induction on m and n to the case $m = n = 1$. In that case ${}_1W_1 = {}_p\alpha_k$, and $\langle\, ,\, \rangle$ is not zero, hence the given homomorphism ${}_p\alpha_k \longrightarrow D({}_p\alpha_k)$ is not zero; but, because ${}_p\alpha_k$ is simple, it is an isomorphism, and the proof is complete.

5. Dieudonné modules (Affine unipotent groups).

From now on, the field k is supposed to be perfect.

Let $\underrightarrow{W}$ be the inductive system of $\underline{Acu}_k$.

$$\underrightarrow{W}: W_{1k} \xrightarrow{\;T\;} W_{2k} \xrightarrow{\;T\;} W_{3k} \xrightarrow{\;T\;} \cdots .$$

The ring $W(k)$ operates on $\underrightarrow{W}$ as follows. First, we denote by $\sigma : a \longmapsto a^{(p)}$ the Frobenius homomorphism $W(k) \longrightarrow W(k)$, and by $a \longmapsto a^{(p^n)}$ its nth power, $n \in \mathbb{Z}$ ($a \longmapsto a^{(p)}$ is bijective, because k is perfect.) Let $a \in W(k)$ and

$w \in W_n(R), R \in \underline{M}_k$; then we define

$$a * w = a^{(p^{1-n})}{}_R \cdot w,$$

where $a^{(p^{1-n})}{}_R$ is the image of $a^{(p^{1-n})}$ in $W(R)$, and $b \cdot w \in W_n(R)$ the product of $b \in W(R)$ and $w \in W_n(R) = W(R)/T^n W(R)$. By this definition, $W_n(R)$ becomes a $W(k)$-module, and $T: W_n(R) \longrightarrow W_{n+1}(R)$ is a homomorphism of $W(k)$-module, because

$$T(a * w) = T(a^{(p^{1-n})}{}_R \cdot w) = T(F(a^{(p^{-n})}{}_k) \cdot w) = a^{p^{-n}} . Tw = a * Tw.$$

For any $G \in \underline{Acu}_k$, we define the <u>Dieudonné module</u> $M(G)$ of G to be the $W(k)$-module

$$M(G) = \varinjlim \underline{Acu}_k(G, W_{nk})$$

(equivalently $M(G) = \text{Ind}\,(\underline{Acu}_k)\,(G, \underrightarrow{W})$). Of course, $G \longmapsto M(G)$ is a contravariant functor from $\underline{Acu}_k$ to category $\underline{\text{Mod}}\ W(k)$ of all $W(k)$-modules. This construction obviously commutes with automorphisms $k \simeq k$, in particular with $f_k : k \longrightarrow k$. If M is a $W(k)$-module, let $M^{(p)} = M \otimes_{W(k),\sigma} W(k)$: as a group $M^{(p)} = M$, but the external law is $(w, m) \longrightarrow w^{(p^{-1})} m$; if $f \in \underline{Acu}_k(G, W_{nk})$, then $f^{(p)}$ is a homomorphism from $G^{(p)}$ to $W_{nk}^{(p)} = W_{nk}$. Hence a map $f \longmapsto f^{(p)}$ from $M(G)$ to $M(G^{(p)})$; it is clear that $(wf)^{(p)} = w^{(p)} f^{(p)}$ for $w \in W(k)$, and this map induces an isomorphism,

$$M(G)^{(p)} \xrightarrow{\sim} M(G^{(p)}),$$

by means of which we always identify $M(G^p)$ with $M(G)^{(p)}$

The two morphisms F_G and V_G define two morphisms $F = M(F_G) : M(G)^{(p)} \longrightarrow M(G)$, and $V = M(V_G) : M(G) \longrightarrow M(G)^{(p)}$ or equivalently group-homomorphisms $F, V : M(G) \longrightarrow M(G)$ with $F(am) = a^{(p)} Fm$, $V(a^{(p)} m) = aVm$,

$a \in W(k)$, $m \in M(G)$. By construction, if $\bar{m} \in \underline{Acu}_k(G, W_{nk})$ represents $m \in M(G)$, Fm and Vm are represented by $F_{W_{nk}} \circ \bar{m}$ and $V_{W_{nk}} \circ \bar{m}$.

The morphism $T: W_{nk} \longrightarrow W_{n+1k}$ being a monomorphism, the maps $\underline{Acu}_k(G, W_{nk}) \longrightarrow \underline{Acu}_k(G, W_{n+1k})$ are injective, and $\underline{Acu}_k(G, W_{nk})$ can be identified with a submodule of $M(G)$; more precisely

$$\underline{Acu}_k(G, W_{nk}) = \{ m \in M(G),\ V^n m = 0 \}.$$

It follows that any element of $M(G)$ is killed by a power of V.

Let D_k be the (non-commutative) ring generated by $W(k)$ and two elements F and V subject to the relations

$$Fw = w^{(p)} F,\ w^{(p)} V = Vw,\ FV = VF = p$$

It can be easily seen that any element of D_k can be written uniquely as a finite sum

$$\sum_{i>0} a_{-i} V^i + a_o + \sum_{i>0} a_i F^i .$$

If $G \in \underline{Acu}_k$, then $M(G)$ has a canonical structure of a left D_k-module; if K is a perfect extension of k, there is a canonical map of D_K-modules

$$(*) \qquad W(K) \otimes_{W(k)} M(G) \longrightarrow M(G \otimes_k K)$$

(remark that $D_K \cong W(K) \otimes_{W(k)} D_k$, and that the left hand side can also be written $D_K \otimes_{D_k} M(G)$).

Theorem. The functor M induces an anti-equivalence between Acu_k and the category of all D_k-modules of V-torsion. For any perfect extension K of k, (*) is an isomorphism. Moreover

$$G \text{ is algebraic} \Longleftrightarrow M(G) \text{ is a finitely generated } D_k\text{-module},$$

$$G \text{ is finite} \Longleftrightarrow M(G) \text{ is a } W(k)\text{-module of finite length.}$$

Proof in D.G. V, § 1, n° 4.

6. Dieudonné modules (p-torsion finite k-groups)

Proposition. The functor $G \longmapsto M(G)$ induces an anti-equivalence between Feu_k (resp. Fiu_k) and the category of D_k-modules, which are W(k)-modules of finite length, killed by a power of V and on which F is bijective (resp. and killed by a power of F).

This follows from the theorem, and the fact that if G is finite, then G is etale (resp. infinitesimal) if and only if F_G is an isomorphism (resp. $F_G^n = 0$ for large n).

Examples. If $G = (\mathbb{Z}/p\mathbb{Z})_k \in Feu_k$, then $M(G) = k$ with $F = 1$, $V = 0$; if $G = {}_p\alpha_k \in Fiu_k$, then $M(G) = k$ with $F = 0$, $V = 0$.

Proof. We can suppose k algebraically closed, in either case G is the unique simple object of Feu_k (resp: Fiu_k); hence M(G) is the unique simple object of the corresponding category) and it is clear that the proposed modules are simple.

Corollary. For $G \in Feu_k$ or Fim_k, we have

$$rk(G) = p^{\text{length } (M(G))}.$$

We can replace k by $\overline{k}$, and it is enough to check the formulas for the simple groups, in which case it follows from the examples above.

Let m,n be two positive integers; consider the canonical injection ${}_mW_n \longrightarrow W_n$; it defines an element $u \in M({}_mW_n)$, clearly $V^n u = F^n u = 0$, hence a map of D-modules $(D = D_k)$:

$$\lambda_{m,n} : D/(DF^m + DV^n) \longrightarrow M({}_mW_n).$$

<u>Proposition</u>. $\lambda_{m,n}$ <u>is bijective</u>.

Using the exact sequences connecting the ${}_mW_n$, we are easily reduced to the case $m = n = 1$; but $D/DF + DV \cong k$ and $M({}_1W_1) = M({}_p\alpha_k) = k$.

Take $m = n$. Any element in $D/(DF^n + DV^n)$ can be written in a unique way $x = w_{1-n} V^{n-1} + \dots + w_{-1}V + w_o + w_1 F + \dots + w_{n-1} F^{n-1}$ where $w_i \in W_{n-|i|}(k)$; we therefore have a canonical $W(k)$-linear projection

$$\pi_n : M({}_nW_n) \longrightarrow W_n(k)$$

defined by $\pi_n(\lambda_n(x)) = w_o$.

Let Q be the quotient field of $W(k)$, and W_∞ be the $W(k)$-module $Q/W(k)$; it can be identified with the direct limit of the system

$$W(k)/_{pW(k)} \xrightarrow{p} W(k)/_{p^2W(k)} \longrightarrow \dots ;$$

but this system is also

$$W_1(k) \xrightarrow{T} W_2(k) \xrightarrow{T} W_3(k) \longrightarrow .$$

Hence $W_\infty = \varinjlim W_n(k) = \underrightarrow{W}(k)$.

For any D_k-module M, we denote by M^* the following D_k-module: as $W(k)$-module, $M^* = \underline{\mathrm{Mod}}_{W(k)}(M, W_\infty)$; if $f \in M^*$, then $(Ff)(m) = f(Vm)^{(p)}$, $(Vf)(m) = f(Fm)^{(p^{-1})}$. It is clear (duality of finite length modules over a principal ideal ring) that $M \longrightarrow M^*$ induces a duality in the category of D_k-modules which are of finite length over $W(k)$.

Let now $G \in \underline{\mathrm{Fiu}}_k$, then there exists n such that $V_G^n = 0$, $F_G^n = 0$; it follows that $M(G) = \underline{\mathrm{Fiu}}_k(G, {}_nW_n)$; moreover $V_{D(G)}^n = 0$, $F_{D(G)}^n = 0$, and $M(D(G)) = \underline{\mathrm{Fiu}}_k(D(G), {}_nW_n)$. Let $m: D(G) \longrightarrow {}_nW_n$ be an element of $M(D(G))$; let $ah_n: {}_nW_n \longrightarrow D({}_nW_n)$ be the isomorphism given in the n° 1, and look at the composed homomorphism

$$ {}_nW_n \xrightarrow{ah_n} D({}_nW_n) \xrightarrow{D(m)} D(DG) \cong G \quad ; $$

this gives a D-linear map

$$ \varphi_m: M(G) \longrightarrow M({}_nW_n); $$

composing this with $\pi_n: M({}_nW_n) \longrightarrow W_n(k)$ and the canonical injection $W_n(k) \longrightarrow W_\infty$, we get a $W(k)$-linear map $M(G) \longrightarrow W_\infty$, i.e. an element of $M(G)^*$. Hence a map

$$ (**) \qquad M(D(G)) \longrightarrow M(G)^*. $$

This map is independent of the choice of the integer n: if we replace $m: D(G) \longrightarrow {}_nW_n$ by $m' = itm = tim: D(G) \longrightarrow {}_{n+1}W_{n+1}$, then $D(m)ah_n$ is replaced by $D(m')ah_{n+1} = D(itm)ah_{n+1} = D(m)D(it)ah_{n+1} = D(m)ah_n fv$; hence φ_m is replaced by $\varphi_{m'} = M(D(m)ah_n\, fv) = M(fv)M(D(m)ah_n) = M(fv)\varphi_m$ and $\pi_n\varphi_m$ is replaced by $\pi_{n+1}M(fv)\varphi_m$. But $M(fv): D/(DF^n + DV^n) \longrightarrow D/(DF^{n+1} + V^{n+1})$ is of course $x \longrightarrow FVx = px$, and $\pi_{n+1}M(fv) = \pi_{n+1}\, p = \pi_n$.

We therefore have a well-defined $W(k)$-linear map (**).

<u>Theorem.</u> <u>For all</u> $G \in \underline{\mathrm{Fiu}}_k$, (**) <u>is an isomorphism of</u> D_k-<u>modules</u>.

The proof runs as follows.

a) (**) commutes with F and V.

b) Theorem is true if $G = {}_nW_n$.

c) Any G is a subgroup of a $({}_nW_n)^r$.

For the details, see D.G. V, § 4, n° 5.

In short, the autoduality $G \longmapsto D(G)$ of $\underline{Fiu}_k$ corresponds, *via* the Dieudonné functor, to the autoduality $M \longmapsto M^*$ in the category of D_k-module of finite length killed by a power of V and F.

Let now $G \in \underline{Fim}_k$, we define the Dieudonné module $M(G)$ by

$$M(G) = M(D(G))^*.$$

It follows from the Cartier duality between $\underline{Fim}_k$ and $\underline{Feu}_k$ that the functor $G \longrightarrow M(G)$ just defined induces an antiequivalence between $\underline{Fim}_k$ and the category of all D_k-modules of finite length on which F is nilpotent and V bijective.

We can describe $M(G)$ as follows. Suppose first G is diagonalisable: $G = D(\Gamma_k)$. Then $D(G) \cong \Gamma_k$, and $M(D(G)) = \varinjlim \underline{Acu}_k(\Gamma_k, W_{nk}) = \varinjlim \underline{Gr}(\Gamma, W_n(k)) = \underline{Gr}(\Gamma, W_\infty) = \underline{Mod}_{W(k)}(W(k) \otimes_{\mathbb{Z}} \Gamma, W_\infty)$, hence

$$M(G) \cong W(k) \otimes_{\mathbb{Z}} \Gamma.$$

In general, G is defined by a Galois module Γ and $M(G)$ is the set of invariants under the Galois group π of $M(G \otimes_k \bar{k})$; hence

$$M(G) \cong (W(\bar{k}) \otimes_{\mathbb{Z}} \Gamma)^{\pi}.$$

Moreover, F and V are easily described by duality:

$$F(\lambda \otimes \chi) = \lambda^{(p)} \otimes p\chi$$

$$V(\lambda \otimes \chi) = \lambda^{(p^{-1})} \otimes \chi.$$

Let $\underline{F}_{p_k}$ be the category of all finite k-groups of p-torsion. Any G in $\underline{F}_{p_k}$ decomposes uniquely as $H \times K$, with $H \in \underline{Fiu}_k \times \underline{Feu}_k, K \in \underline{Fim}_k$ and we define $M(G)$ as $M(H) \times M(K)$.

Theorem a) The functor $G \mapsto M(G)$ is an antiequivalence between the category $\underline{Fp}_k = \underline{Fiu}_k \times \underline{Feu}_k \times \underline{Fim}_k$ of all finite k-groups of p-torsion, and the category of all triples (M, F_M, V_M) where M is a finite length $W(k)$-module and F_M and V_M two group endomorphisms of M such that

$$F_M(\lambda m) = \lambda^{(p)} F_M(m)$$

$$V_M(\lambda^{(p)} m) = \lambda V_M(m)$$

$$F_M V_M = V_M F_M = p.\mathrm{id}_M.$$

b) G is etale, infinitesimal, unipotent or multiplicative according as F_M is isomorphic, F_M nilpotent, V_M nilpotent, or V_M isomorphic.

c) For any $G \in \underline{Fp}_k$, one has

$$rk(G) = p^{\text{length } M(G)}.$$

d) If K is a perfect extension of k, there exists a functorial isomorphism

$$M(G \otimes_k K) \cong W(K) \otimes_{W(k)} M(G).$$

e) There exists a functorial isomorphism

$$M(D(G)) = M(G)^*.$$

Let (M, F_M, V_M) be as in the theorem. There exists m with $F_M^{2m} M = F_M^m M$, then $M = \mathrm{Ker}\, F_M^m \oplus \mathrm{Im}\, F_M^m = M_0 \oplus M_1$ where M_0, M_1 are stable by F and V, $F^m M_0 = 0$ and $F|M_1$ is bijective; similarly $M_0 = M_{00} \oplus M_{01}, M_1 = M_{10} \oplus M_{11}$, with $V^n M_{00} = 0$, $V^n M_{10} = 0$, $V|M_{01}$ is bijective, $V|M_{11}$ is bijective. But $FV = VF = p$, hence $M_{11} = 0$; this implies $M = M_{00} \oplus M_{01} \oplus M_{10}$.

The proof is now straight forward and left as an exercise.

8. Dieudonne modules (p-divisible groups).

Let us first prove a lemma.

Lemma. *Let* $\cdots \longrightarrow M_{n+1} \xrightarrow{\pi_n} M_n \longrightarrow \cdots \longrightarrow M_1$ *be a system of* $W(k)$*-modules with the following properties.*

1) *The sequence* $M_{n+1} \xrightarrow{p^n} M_{n+1} \xrightarrow{\pi_n} M_n \longrightarrow 0$ *is exact for all* n.

2) M_n *is of finite length for all* n.

Let $M = \varprojlim M_n$. *Then* M *is a finitely generated* $W(k)$*-module and the canonical map* $M \longrightarrow M_n$ *identifies* M_n *with* M/p^nM, *for all* n.

It follows from 1) that

$$M_{n+m} \xrightarrow{p^n} M_{n+m} \xrightarrow{\pi} M_n \longrightarrow 0$$

is exact for all n and m (where $\pi = \pi_n \circ \pi_{n+1} \circ \cdots \circ \pi_{m-1}$). Taking the inverse limit over m, we find an exact sequence

$$M \xrightarrow{p^n} M \xrightarrow{\lambda_n} M_n \longrightarrow 0$$

[the $\varprojlim$ functor is exact for finite length modules - D.G. V § 2, 2.2 a)] where λ_n is the canonical projection, hence the last assertion. Let now $m_1, \ldots, m_r$ be elements in M generating $M/pM = M_1$; consider the $W(k)$-module homomorphism $\varphi : W(k)^r \longrightarrow M$ such that $\varphi(a_1, \ldots, a_r) = a_1 m_1 + \cdots + a_r m_r$. It induces surjective maps $W(k)^r/p^n W(k)^r \longrightarrow M/p^n M$ for all n hence is surjective as an inverse limit of surjective maps of finite length modules.

Alternative proof. Apply Bourbaki, *Alg. Com.*, Ch. 3, § 2, n°.11, Prop. 14 and Cor. 1 to $A_i = W(k)/p^{i+1}W(k)$, $M_i = M_{i+1}$.

We say that a formal group G is of p-torsion if

1) $G = \bigcup \operatorname{Ker} p^n \operatorname{id}_G$

2) $\operatorname{Ker} p \operatorname{id}_G$ is finite.

We have exact sequences

$$0 \longrightarrow \operatorname{Ker} p^n \longrightarrow \operatorname{Ker} p^{n+1} \xrightarrow{p^n} \operatorname{Ker} p^{n+1}$$

$$0 \longrightarrow \operatorname{Ker} p^n \longrightarrow \operatorname{Ker} p^{m+n} \xrightarrow{p^n} \operatorname{Ker} p^m$$

the latter show by induction that $\operatorname{Ker} p^n$ is finite for all n. Define $M(G) = \varinjlim M(\operatorname{Ker} p^n)$.

Theorem. $G \longrightarrow M(G)$ is an antiequivalence between the category of p-torsion formal groups and the category of triples (M, F_M, V_M) where M is a finitely generated $W(k)$-module and F_M, V_M two group endomorphisms of M with

$$F_M(wm) = w^{(p)} F_M(m)$$

$$V_M(w^{(p)} m) = w V_M(m)$$

$$F_M V_M = V_M F_M = p \operatorname{id}_M .$$

It follows from the lemma that $M(G)$ is finitely generated and that $M_n \cong M(G)/p^n M(G)$. Conversely if M is as before, then we define G as $\varinjlim G_n$ where $M(G_n) = M/p^n M$.

From the definitions and what was already proved follow immediately:

1) G is finite if and only if $M(G)$ is finite, and in that case $M(G)$ is the same as defined in § 7.

2) G is p-divisible if and only if $M(G)$ is torsion-less (= free), and

$$\text{height}\,(G) = \dim M(G),$$

3) <u>For any perfect extension</u> K/k, <u>there is a functorial isomorphism</u>

$$M(G \otimes_k K) \cong W(K) \underset{W(k)}{\otimes} M(G).$$

4) <u>If</u> G <u>is</u> p-<u>divisible, with Serre dual</u> G', <u>then</u>

$$M(G') = \mathrm{Mod}_{W(k)}(M(G), W(k)),$$

<u>with</u> $(F_{M(G')}f)(m) = f(V_M m)^{(p)}$, $(V_{M(G')}f)(m) = f(F_M m)^{(p^{-1})}$.

<u>Proof of</u> 4. Let $M(G) = M$; then $M = \varprojlim M/p^nM$, and $M/p^nM = M(\mathrm{Ker}\ p^n\ \mathrm{id}_G)$; but G' is defined as $\varinjlim D(\mathrm{Ker}\ p^n\ \mathrm{id}_G)$, hence $M(G') = \varprojlim M(D(\mathrm{Ker}\ p^n\ \mathrm{id}_G)) =$

$$\varprojlim (M/p^nM)^* = \varprojlim \mathrm{Mod}_{W(k)}(M/p^nM, W(k)/p^nW(k)) = \mathrm{Mod}_{W(k)}(M, W(k)).$$

9. <u>Dieudonné modules (connected formal group of finite type)</u>.

By a similar discussion (replacing p by F), we have the following results: if G is a connected finite type formal group, define $M(G) = \varprojlim M(\mathrm{Ker}\ F^n_G)$; it is a module over the F-completion $\hat{D}_k$ of D_k.

<u>Theorem</u>. $G \longrightarrow M(G)$ <u>is an antiequivalence between the category of connected formal groups of finite type and the category of finite type</u> $\hat{D}_k$-<u>modules</u> M <u>such that</u> M/FM <u>has finite length. Moreover</u>

1) G <u>finite</u> $\Longleftrightarrow$ $M(G)$ <u>has finite length</u> $\Longleftrightarrow$ $F^nM(G) = 0$ <u>for</u> n <u>large</u>.

2) G <u>smooth</u> $\Longleftrightarrow$ $F: M(G) \longrightarrow M(G)$ <u>is injective; in that case</u>

$\dim(G) = \mathrm{length}\big(M(G)/FM(G)\big)$.

CHAPTER IV

CLASSIFICATION OF p-DIVISIBLE GROUPS

k is a perfect field (unless otherwise stated), charac (k) $\neq 0$; we denote by $B(k)$ the quotient field of $W(k)$, and extend $x \mapsto x^{(p)}$ to an automorphism of $B(k)$; the set of fixed points of $x \mapsto x^{(p)}$ in $W(k)$ (resp. $B(k)$) is $W(\mathbb{F}_p) = \mathbb{Z}_p$ (resp. $B(\mathbb{F}_p) = \mathbb{Q}_p$).

1. Isogenies.

A F-lattice (resp. F-space) over k is a free $W(k)$-module (resp. a $B(k)$-vector space), of finite rank, together with an injective (resp. injective = bijective) group endomorphism F such that $F(\lambda x) = \lambda^{(p)}Fx$. If M is a F-lattice, then $B(k) \otimes_{W(k)} M$ has a natural F-space structure.

To each p-divisible group G, we associate the F-lattice $M(G)$, and the F-space $E(G) = B(k) \otimes_{W(k)} M(G)$; the functor $G \longrightarrow M(G)$ is an antiequivalence between p-divisible groups and those F-lattices M for which $FM \supset pM$.

If K is a perfect extension of k, and M a F-lattice over k, we define M_K as $W(K) \otimes_{W(k)} M$, similarly for F-spaces.

Lemma. Let G and H be two p-divisible groups of the same height and $f: G \longrightarrow H$ be a homomorphism. The following conditions are equivalent

a) Ker f is finite,

b) f is an epimorphism,

c) $M(f): M(H) \longrightarrow M(G)$ is injective,

d) Coker $M(f)$ is finite,

e) $E(f): E(H) \longrightarrow E(G)$ is an isomorphism.

This is clear: (a) $\Longleftrightarrow$ (d), (b) $\Longleftrightarrow$ (c), and (c) $\Longleftrightarrow$ (d) $\Longleftrightarrow$ (e). Such an f is called an isogeny.

Proposition. Let G and H be two p-divisible groups. Then $E(G)$ and $E(H)$ are isomorphic if and only if there exists an isogeny $f:G \longrightarrow H$.

Let $\varphi:E(H) \longrightarrow E(G)$ be an isomorphism; there exists m such that $\varphi(M(H)) \subset p^{-m}M(G)$, then $p^m\varphi:M(H) \longrightarrow M(G)$ corresponds to an isogeny f. The converse is clear.

Two such groups are called isogenous. The classification of p-divisible groups upto isogeny is therefore equivalent to classification of F-spaces of the form $E(G)$.

A F-space E is called effective if it contains a lattice (i.e. a $W(k)$-submodule M such that $E = B(k)\otimes_{W(k)} M$) stable by F, i.e. if it comes from an F-lattice. It comes from a p-divisible group if and only if it contains a lattice stable by F and pF^{-1}.

2. The category of F-spaces

It is a $\mathbb{Q}_p$-linear category: an abelian category, such that $\mathrm{Hom}(E_1,E_2)$ has a natural (finite dimensional, in fact) $\mathbb{Q}_p$-vector space structure, the composition map $(f,g) \longrightarrow g \circ f$ being $\mathbb{Q}_p$-bilinear [note that $\mathbb{Q}_p$ is the centre of $B(k)$].

It has tensor products and internal Hom: If E_1, E_2 are F-spaces, then $E_1 \otimes E_2$ and $\underline{\mathrm{Hom}}(E_1,E_2)$ are the usual $\otimes$ and Hom of $B(k)$-vector spaces and $F(x\otimes y) = Fx \otimes Fy$, $(Fu)(x) = u(F^{-1}x)^{(p)}$, $x \in E_1, y \in E_2, u \in \underline{\mathrm{Hom}}(E_1,E_2)$.

We denote by $\mathbf{1}$ the F-space $(B(k), x \longrightarrow x^{(p)})$, by $\mathbf{1}(n)$ the F-space $B(k), x \longrightarrow p^{-n}x^{(p)}$. The dual $\check{E}$ of E is $\underline{\mathrm{Hom}}(E, \mathbf{1})$, the nth twist $E(n)$ of E is $E\otimes\mathbf{1}(n)$.

We have the usual canonical isomorphisms

$$\mathrm{Hom}(A, \underline{\mathrm{Hom}}(B,C)) = \mathrm{Hom}(A\otimes B,C)$$

$$\underline{\mathrm{Hom}}(\mathbf{1},A) = A$$

$$\mathrm{Hom}(A,B) = \mathrm{Hom}(\mathbf{1},\underline{\mathrm{Hom}}(A,B))$$

$$A\otimes(B\otimes C) = (A\otimes B)\otimes C \ldots$$

In particular

$$E(m)(n) = E(m+n)$$

$$\check{E}(m) = \check{E}(-m).$$

If G is a p-divisible group and G' its Serre dual, then

$$E(G') = \underline{\mathrm{Hom}}(E(G),\mathbf{1}(-1)) = \check{E}(G)(-1)$$

(because Serre duality sends F to $V = pF^{-1}$).

These constructions commute with the base-extension functor $E \longmapsto E_K = B(K)\otimes_{B(k)}E$ $[K/k$ a perfect extension$]$.

3. The F-spaces E^{λ}, $\lambda \geq 0$.

(A) Let $\lambda \geq 0$ be a rational number; write $\lambda = \frac{s}{r}$, with $r,s\in \mathbb{N}$, $r>0$, $(r,s) = 1$. We define the F-lattice M^{λ} over $\mathbb{F}_p$ by

$$M^{\lambda} = \mathbb{Z}_p[T]/(T^r - p^s),$$

F acting by multiplication by T, and similarly, the F-space E^{λ} over $\mathbb{F}_p$ by

$$E^{\lambda} = \mathbb{Q}_p[T]/(T^r - p^s);$$

If $0 \leqslant \lambda \leqslant 1$, then $r \geqslant s$; define $\bar{M}^{\lambda} = \mathbb{Z}_p[F]/(F^{r-s} - V^s)$, then $\bar{M}^{\lambda}$ is a lattice in E^{λ} and a Dieudonné module; actually, let G^{λ} be the p-divisible group over $\mathbb{F}_p$ defined by the exact sequence

$$0 \longrightarrow G^{\lambda} \longrightarrow W(p) \xrightarrow{F^r - V^s} W(p)$$

where $W(p) = \varinjlim (\operatorname{Ker} p^n : W_{\mathbb{F}_p} \longrightarrow W_{\mathbb{F}_p})$. It is clear that

$$M(G^{\lambda}) \simeq \bar{M}^{\lambda}, \; E(G^{\lambda}) \simeq E^{\lambda}.$$

Hence height $(G^{\lambda}) = r$, dim $(G^{\lambda}) = s$. It is also clear that $(G_{\lambda})' = G_{1-\lambda}$.

We put $E_k^{\lambda} = (E^{\lambda})_k = B(k) \otimes_{\mathbb{Q}_p} E^{\lambda}$. It has a $B(k)$-basis $e_1, \ldots, e_r$ $[e_i$ = class of $T^{i-1}]$ such that, if $x = \sum a_i e_i$, then

$$Fx = p^s a_n^{(p)} e_1 + a_1^{(p)} e_2 + \ldots + a_{n-1}^{(p)} e_n.$$

In particular

$$(F^r - p^s)(x) = p^s\left[(a_1^{(p)} - a_1)e_1 + \ldots + (a_n^{(p)} - a_n)e_n\right].$$

(B) Let $W(k)(p^{1/r})$ and $B(k)(p^{1/r})$ be defined by

$$W(k)(p^{1/r}) = W(k)[X]/(X^r - p), B(k)(p^{1/r}) = B(k)[X]/(X^r - p);$$

denote the class of X by $p^{1/r}$, then $W(k)(p^{1/r})$ is a complete descrete valuation ring with residue field k and maximal ideal generated by $p^{1/r}$. We extend $x \longmapsto x^{(p)}$ to $W(k)(p^{1/r})$ and $B(k)(p^{1/r})$ by putting $(p^{1/r})^{(p)} = p^{1/r}$.

Let $F_s : W(k)(p^{1/r}) \longrightarrow W(k)(p^{1/r})$ be defined by

$$F_s(\sum w_i p^{i/r}) = \sum w_i^{(p)} p^{(s+i)/r}$$

and similarly for $B(k)(p^{1/r})$. <u>Then the</u> F-<u>lattice</u> $(W(k)(p^{1/r}), F_s)$ <u>is isomorphic to</u> M_k^λ, <u>the</u> F-<u>space</u> $(B(k)(p^{1/r}), F_s)$ <u>isomorphic to</u> E_k.

<u>Proof.</u> Send $p^{i/r}$ to the class of T^i.

Ⓒ Let $a, b \in \mathbb{N}$ be such that $ar - bs = 1$. Consider $B(\mathbb{F}_{p^r})$ [it is the unique unramified extension of degree r of $B(\mathbb{F}_p) = \mathbb{Q}_p$] and let K^λ be the associative $B(\mathbb{F}_{p^r})$-algebra with unit generated by an element ξ such that

$$\xi^r = p, \ \xi\alpha = \alpha^{(p^{-b})}\xi, \ \alpha \in B(\mathbb{F}_{p^r}).$$

It is a left vector space of dimension r over $B(\mathbb{F}_{p^r})$ with basis $1, \dots, \xi^{r-1}$, hence an <u>algebra of</u> degree r^2 over $\mathbb{Q}_p$. Moreover, because $-b$ is invertible modulo r, $\alpha^{(p^{-b})} = \alpha$ implies $\alpha \in \mathbb{Q}_p$, and K^λ has <u>centre</u> $\mathbb{Q}_p$. Finally, K^λ is a <u>division-algebra</u>: let $X = \sum_{i=0}^{r-1} a_i \otimes \xi^i$ be a right zero divisor. By multiplication by suitable powers of p and ξ, we can suppose that $a_i \in W(\mathbb{F}_{p^i})$, and $a_0 \notin pW(\mathbb{F}_{p^r})$. The matrix of the right multiplication by X in the basis $1, \dots, \xi^{i-1}$ is (write σ for (p^{-b}))

$$\begin{pmatrix} a_0 & a_2 & \cdots & a_{r-1} \\ pa^\sigma_{r-1} & a^\sigma_o & \cdots & a^\sigma_{i-2} \\ & & \cdots & \\ pa^{\sigma^{r-1}}_{i-1} & \cdots & \cdots & a_0^{\sigma^{r-1}} \end{pmatrix}$$

Its determinant is congruent to $a_0 a_0^{\sigma} \ldots a_0^{\sigma^{r-1}} \equiv \mathrm{Norm}(a_0) \bmod p$; it therefore cannot be zero, contradiction.

<u>Suppose now</u> $k \supset \mathbb{F}_{p^r}$, and consider

$$W(k) \otimes_{W(\mathbb{F}_{p^r})} K^{\lambda} = B(k) \otimes_{B(\mathbb{F}_{p^r})} K^{\lambda}.$$

It is a $B(k)$-vector space with basis $\xi^i = 1 \otimes \xi^i$, $i = 0,\ldots,r-1$ and a right K^{λ}-vector space; we make it a F-space over k by defining $F\xi^i = \xi^{i+s}$.

<u>Proposition</u>. a) <u>The F-space</u> $B(k) \otimes_{B(\mathbb{F}_{p^r})} K^{\lambda}$ <u>is isomorphic to</u> E_k^{λ}.

b) <u>Its endomorphisms are the right multiplication by elements of</u> K^{λ}.

We send the F-space $E = B(k) \otimes_{B(\mathbb{F}_{p^r})} K^{\lambda}$ to $B(k)(p^{1/r})$ by mapping ξ^i to $p^{i/r}$; it is easy to check that this mapping is an isomorphism of F-spaces hence a). To prove b), we first remark that the F-space structure and the k^{λ}-vector space structure on E commute: each multiplication $x \longrightarrow x\alpha$, $\alpha \in K^{\lambda}$ is a F-space endomorphism. We use now the following lemma.

<u>Lemma</u>. <u>Let</u> H <u>be any F-space over</u> k; <u>the map</u> $\varphi \longrightarrow \varphi(e_1)$ <u>is a bijection from</u> $\mathrm{Hom}(E_k^{\lambda}, H)$ <u>to the set of all</u> x <u>in</u> H <u>such that</u> $F^r x = p^s x$.

This is clear from the definition of E_k^{λ}.

Using this lemma, it is enough to prove that the elements x of E with $F^r x = p^s x$ are the $1 \otimes \alpha$, $\alpha \in K^{\lambda}$. Let

$$x = \sum_{i=0}^{r-1} \alpha_i \otimes \xi^i, \alpha_i \in B(k);$$

then $F^r x = \sum p^s \alpha_i^{(p^r)} \otimes \xi^i$, and $F^r x = p^s x$ implies $\alpha_i^{(p^r)} = \alpha_i$, i.e.

$\alpha_i \in B(\mathbb{F}_{p^r})$, i.e. $x = 1 \otimes \sum \alpha_i \xi^i \in 1 \otimes K^\lambda$.

(D) Let $\lambda' = s'/r'$, with $r', s' \in \mathbb{N}, (s', r') = 1$, be another positive rational.

<u>Proposition.</u> a) <u>If</u> $\lambda \neq \lambda'$, <u>then</u> $\mathrm{Hom}(E_k^\lambda, E_k^{\lambda'}) = 0$,

b) <u>let</u> $m = $ g.c.d. (r, r'), then

$$E_k^\lambda \otimes E_k^{\lambda'} = (E_k^{\lambda+\lambda'})^m,$$

$$K^\lambda \otimes_{\mathbb{Q}_p} K^{\lambda'} \simeq M_m(K^{\lambda+\lambda'}).$$

a) By the above lemma, we have to look to those $x \in E_k^{\lambda'}$ with $(F^r - p^s)x = 0$; but $E_k^{\lambda'}$ has a basis f_j such that, if $x = \sum b_j f_j$, then $F^{r'} x = \sum b_j^{(p^{r'})} p^{s'} f_j$, hence $F^{rr'} x = \sum b_j^{(p^{rr'})} p^{s'r} f_j$; on the other hand, if $F^r x = p^s x$, then $F^{rr'} x = p^{sr'} x = \sum b_j p^{sr'} f_j$. Because $sr' \neq s'r$, and $v(\lambda^{(p)}) = v(\lambda)$ for $\lambda \in B(k)$, this implies $x = 0$.

b) Let $e'_1, \dots e'_{r'}$ be the canonical base of $E_k^{\lambda'}$, and $\lambda + \lambda' = \lambda_o = s_o/r_o$, with $s_o = sr' + r's/m$, $r_o = rr'/m$. Then

$$F^{r_o}(e_i \otimes e'_j) = F^{\frac{r'}{m} r} e_i \otimes F^{\frac{r}{m} r'} e'_j = p^{\frac{r'}{m} s} e_i \otimes p^{\frac{r}{m} s'} e'_j = p^{s_o}(e_i \otimes e'_j).$$

It follows that, i and j being fixed, and indices running modulo (r, r'), the vectors $e_{i+k} \otimes e'_{j+k}$, $k = 0, \dots, r_o - 1$, span a sub-F-space of $E_k^\lambda \otimes E_k^{\lambda'}$ isomorphic to $E_k^{\lambda+\lambda'}$. This gives m linearly independent subspaces, hence an isomorphism $E_k^\lambda \otimes E_k^{\lambda'} \simeq (E_k^{\lambda+\lambda'})^m$.

Taking k big enough, this gives a map of the endomorphism algebras

$$K^{\lambda}\otimes_{\mathbb{Q}_p} K^{\lambda'} \longrightarrow M_m(K^{\lambda+\lambda'}),$$

this map is injective because $K^{\lambda}\otimes_{\mathbb{Q}_p} K^{\lambda'}$ is simple, hence bijective because both sides have dimension $(rr')^2$ over $\mathbb{Q}_p$. As a <u>corollary</u>, take $\lambda' = n\in\mathbb{N}$ in c); we find isomorphisms

$$M_k^{\lambda}(-n) \simeq M_k^{\lambda+n}.$$

(In particular $\mathbb{1}(-n) = M_k^n$), and

$$K^{\lambda} \simeq K^{\lambda+n}.$$

Hence $\lambda\mapsto K^{\lambda}$ gives a <u>homomorphism</u>

$$\mathbb{Q}/\mathbb{Z} \longrightarrow Br(\mathbb{Q}_p),$$

which is <u>injective</u> (because K^{λ} is a <u>skew-field</u>, hence cannot be split if $r \neq 1$, i.e. $\lambda\notin\mathbb{Z}$), and known to be <u>surjective</u>.

Ⓔ For $\lambda\in\mathbb{Q}$, $\lambda\leq 0$ we define E_k^{λ} to be the dual of $E_k^{-\lambda}$ (note that $E^0 = \mathbb{1}$). From the relations between dual, tensor products, and internal Hom, and using the twist operation we obtain for $\lambda, \lambda'\in\mathbb{Q}$

a) $E_k^{\lambda}\otimes E_k^{\lambda'} \simeq (E_k^{\lambda+\lambda'})^m$, $m = \text{g.c.d}(r,r')$,

b) $\underline{\text{Hom}}(E_k^{\lambda}, E_k^{\lambda'}) \simeq (E_k^{\lambda'-\lambda})^m$, $m = \text{g.c.d}(r,r')$,

c) $E_k^{\lambda}(n) = E_k^{\lambda-n}$, $(E_k^{\lambda})^{\vee} = E_k^{-\lambda}$,

d) If $\lambda = \frac{s}{r}$, $r>0$, $(s,r) = 1$, then $\dim E_k^{\lambda} = r$. If $k\supset\mathbb{F}_{p^r}$ then $\text{End}(E_k^{\lambda})$ is a central division algebra over $\mathbb{Q}_p$, with invariant $\lambda \bmod 1$. E_k^{λ} is effective if and only if $\lambda \geq 0$, E_k^{λ} comes from a p-divisible group if and only if $0 \leq \lambda \leq 1$.

e) $$\mathrm{Hom}(E_k^{\lambda}, E_k^{\lambda'}) = 0 \text{ if } \lambda \neq \lambda'.$$

4. Classification of F-spaces over an algebraically closed field.

Lemma 1. *If k is algebraically closed, any extension of E_k^{λ} by $E_k^{\lambda'}$, $\lambda, \lambda' \in \mathbb{Q}$, splits.*

Let $0 \longrightarrow E_k^{\lambda'} \longrightarrow E \xrightarrow{\varphi} E_k^{\lambda} \longrightarrow 0$ be an exact sequence of F-spaces; for any n, we have an exact sequence

$$0 \longrightarrow E_k^{\lambda'+n} \longrightarrow E(-n) \longrightarrow E_k^{\lambda+n} \longrightarrow 0$$

that splits if and only if the first one splits; taking n large enough, we can therefore suppose $\lambda, \lambda' \geq 0$. Write $\lambda = s/r$, $\lambda' = s'/r'$ as usual. It is sufficient to prove

(*) $$F^r - p^s : E_k^{\lambda'} \longrightarrow E_k^{\lambda'} \text{ is surjective.}$$

Indeed, let $x \in E$ be such that $\varphi(x) = e_1$; then $(F^r - p^s)(x) \in E_k^{\lambda'}$. If (*) is true, there exists a $y \in E_k^{\lambda'}$ with $(F^r - p^s)(y) = (F^r - p^s)(x)$. Replacing x by x - y, we can suppose $(F^r - p^s)(x) = 0$, and x gives a splitting.

We have $(F^r - p^s)(F^{r(r'-1)} + F^{r(r'-2)}p^s + \dots + p^{s(r'-1)}) = F^{rr'} - p^{sr'}$, and it is enough to show that $F^{rr'} - p^{sr'} : E_k^{\lambda'} \longrightarrow E_k^{\lambda'}$ is surjective. If $e'_1, \dots, e'_{r'}$ is the canonical basis of $E_k^{\lambda'}$, we have

$$(F^{rr'} - p^{sr'})(\sum a_i e'_i) = \sum (p^{rs'} a_i^{(p^{r'})} - p^{sr'} a_i) e'_i ;$$

it is therefore sufficient to show that, if $\alpha, \beta \in \mathbb{Z}$, the map

$$x \longmapsto p^{\beta} x^{(p^{\alpha})} - x$$

from B(k) to B(k) is surjective.

If $\beta > 0$ then, taking $x = \sum_{i=0}^{\infty} p^{i\beta} b^{(p^{i\alpha})}$, we find

$$p^{\beta} x^{(p^{\alpha})} - x = - b.$$

If $\beta < 0$, we write $p^{\beta} x^{(p^{\alpha})} - x = p^{\beta} x^{(p)} - p^{-\beta} (p^{\beta} x^{(p^{\alpha})})^{(p^{-\alpha})}$, and are reduced to the preceeding case. If $\beta = 0$, we use successive approximation: let $b \in B(k)$ be fixed, and suppose $x \in B(k)$ and $m \in \mathbb{Z}$ are such that $x^{(p^{\alpha})} - x - b \in p^{m} W(k)$; if $x_1 = x + p^{m} y$, $y \in W(k)$, then $x_1^{(p^{\alpha})} - x_1^{(p^{\alpha})} - b = p^{m}(y^{(p^{\alpha})} - y + (x^{(p)} - x - b)/p^{m})$, and this belongs to $p^{m+1} W(k)$ if and only if $y^{-p^{\alpha}} - \bar{y} + \overline{(x^{(p^{\alpha})} - x - b)/p^{m}} = 0$, denoting by $z \longrightarrow \bar{z}$ the residue map $W(k) \longrightarrow k$. Because k is algebraically closed, this equation has a solution.

<u>Lemma 2</u>. <u>Let</u> $F^{n} + a_1 F^{n-1} + \ldots + a_n \in W(k)[F]$ (<u>non-commutative polynomial ring</u>) k <u>algebraically closed</u>. <u>There exists</u> $r, s \in \mathbb{N}$, <u>coprime</u>, <u>and elements</u> $b_0, b_1, \ldots, b_{n-1}$, $u \in W(k)(p^{1/r})$, <u>with</u> u <u>invertible</u>, <u>such that</u>, <u>in</u> $W(k)(p^{1/r})[F]$, <u>we have</u>

$$(**) \qquad F^{n} + a_1 F^{n-1} + \ldots + a_n = (b_0 F^{n-1} + b_1 F^{n-2} + \ldots + b_{n-1})(F - p^{s/r})u.$$

Let $\lambda = \inf \left(\frac{v(a_i)}{i}\right)$; write $\lambda = s/r$, s and r coprime, and put $a_i = p^{is/r} \alpha_i$; then $\alpha_i \in W(k)$, and α_i is unit for at least one $i > 0$. Let us look for b_i of the form $p^{is/r} \beta_i$, $\beta_i \in W(k)$. Putting $v = u^{-1}$, we can write (**) as:

$$v^{(p^n)} F^n + v^{(p^{n-1})} a_1 F^{n-1} + \dots + va_n =$$

$$b_0 F^n + (b_1 - p^{s/r} b_0) F^{n-1} + \dots + (b_{n-1} - p^{s/r} b_{n-2}) F - p^{s/r} b_{n-1},$$

so that (**) is equivalent to

$$v^{(p^n)} = b_0$$

$$a_1 v^{(p^{n-1})} = b_1 - p^{s/r} b_0$$

$$\dots \qquad \dots \qquad \dots$$

$$\dots \qquad \dots \qquad \dots$$

$$a_{n-1} v^{(p)} = b_{n-1} - p^{s/2} b_{n-2}$$

$$a_n v = -p^{s/r} b_{n-1}.$$

Replacing a_i by $p^{is/r} \alpha_i$ and b_i by $p^{is/r} \beta_i$, we find the system

$$v^{(p^n)} = b_0$$

$$\alpha_1 v^{(p^{n-1})} = b_1 - b_0$$

$$\alpha_{n-1} v^{(p)} = b_{n-1} - b_{n-2}$$

$$\alpha_n v = -b_{n-1}$$

and we have a solution if and only if we can find a unit v in $W(k)(p^{1/r})$ such that

$$v^{(p^n)} + \alpha_1 u^{(p^{n-1})} + \dots + \alpha_n v = 0.$$

This equation, we solve by successive approximation. Modulo $p^{1/r}$, it gives

$$\bar{v}^{p^n} + \bar{\alpha}_1 \bar{v}^{p^{n-1}} + \ldots + \bar{\alpha}_n \bar{v} = 0,$$

and this has a <u>non-zero solution</u> because one of the $\bar{\alpha}_i$ is non-zero and k is algebraically closed; we can therefore start the induction and suppose we have a <u>unit</u> $v_i \in W(k)$ with

$$v_i^{(p^n)} + \alpha_1 v_i^{(p^{n-1})} + \ldots + \alpha_n v_i \equiv 0 \bmod p^{i/r}.$$

Writing $v_{i+1} = v_i + p^{i/r} x$, and solving

$$v_{i+1}^{(p^n)} + \alpha_1 v_{i+1}^{(p^{n-1})} + \ldots + \alpha_n v_{i+1} \equiv 0 \bmod p^{(i+1)/r},$$

we find an equation

$$\bar{x}^{p^n} + \bar{\alpha}_1 \bar{x}^{p^{n-1}} + \ldots + \bar{\alpha}_n \bar{x} = z$$

which has a solution in k.

<u>Lemma 3</u>. <u>Let</u> k <u>be algebraically closed, and let</u> E <u>be a non-zero</u> F-<u>space</u>. <u>There exists a</u> $\lambda \in \mathbb{Q}$ <u>and a non-zero morphism</u> $E \longrightarrow E_k^{\lambda}$.

Taking a non-zero simple quotient of E, we can suppose E simple, i.e. a simple $B(k)[F]$-module. But $B(k)[F]$ is an (non-commutative) euclidean ring, and such a module is a quotient $B(k)[F]/P = B(k)[F]/B(k)[F]P$ where $P \in B(k)[F]$ is a monic polynomial $F^n + a_1 F^{n-1} + \ldots + a_n$. Replacing E by an $E(-m)$, m large, we replace F by $p^m F$, and we can suppose $a_i \in W(k)$. Hence E is defined by the F-lattice $M = W(k)[F]/P$. Then, by lemma 2, we can write $P = Q(F - p^{s/r})u$, where $Q \in W(k)(p^{1/r})[F]$, $u \in W(k)(p^{1/r})^*$, and $(r,s) = 1$. Then $x \longrightarrow xu^{-1}$ gives an epimorphism

$$W(k)(p^{1/r}) \otimes_{W(k)} M \longrightarrow W(k)(p^{1/r})[F]/(F - p^{s/r});$$

but, as a $W(k)[F]$-module, the right-hand side is M_k^λ, and the induced map

$$M \longrightarrow W(k)(p^{1/r}) \otimes_{W(k)} M \longrightarrow M_k^\lambda$$

is a non-zero F-lattice homomorphism.

Proposition. Each E_k^λ is a simple F-space (i.e. does not contain any proper non-zero F-subspace).

We can suppose k algebraically closed. If E is a proper F-subspace of M_k^λ, there exist (lemma 3) a non-zero morphism

$$E_k^\lambda / E \longrightarrow E_k^\mu .$$

If $\mu \neq \lambda$, the composite map $E_k^\lambda \longrightarrow E_k^\mu$ is zero by section 3, E) e) hence $\lambda = \mu$; then this composite map must be an isomorphism, because $\mathrm{End}(E_k^\lambda)$ is a skew-field; this gives $E = 0$.

Theorem (Manin). If k is algebraically closed, the category of F-spaces over k is semi-simple, its simple objects being the E_k^λ; any F-space is isomorphic to a direct sum $\sum (E_k^\lambda)^{m_\lambda}$.

By lemma 3 and the above proposition, the simple F-spaces are just the E_k^λ; by the proposition, any F-space is an extension of E_k^λ. By lemma 1, such an extension splits.

Corollary. If k is algebraically closed, any F-space over k is isomorphic to an F-space E_k, E an F-space over the prime field.

Corollary. If k is algebraically closed, any p-divisible group over k is isogeneous to a product of G_k^λ.

5. Slopes.

Let E be an F-space over k, k algebraically closed. Let $\lambda \in \mathbb{Q}$. The component of slope λ in E is the sum of the sub F-spaces of E isomorphic to E_k; the multiplicity of the slope λ is the B(k)-dimension of this component (e.g., if $\lambda = s/r$, the multiplicity of λ in E_k^λ is r).

The slope-sequence of E is the non-decreasing sequence

$$\lambda_1 \leq \lambda_2 \quad \cdots \leq \lambda_n$$

($n = [E:B(k)]$) of all slopes of E, each one repeated according to its multiplicity.

The Newton polygon P of E is the polygon $OA_1 \ldots A_n$ in $\mathbb{Q}^2$, where A_i has coordinates $(i, \lambda_1 + \ldots + \lambda_i)$; the extremal points of P have integral coordinates and the slopes of its sides are the λ_i.

The slope-function ω of E is the function $\omega : \mathbb{Q} \longrightarrow \mathbb{Q}$ defined by

$$\omega(\lambda) = \sum_{i=1}^{n} \inf(\lambda_i, \lambda)$$

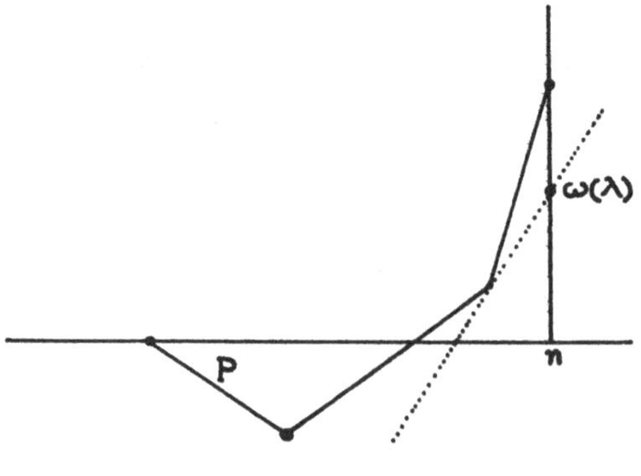

Each of these three objects determine the two others and determine E upto isomorphism; for instance the set above P is

$$\{(x,y) \mid y \geq \lambda(x-n) + \omega(\lambda), \forall \lambda\}.$$

Proposition. *Let* M *be an* F-*lattice, and* ω *the slope function of* $B(k)\otimes_{W(k)}M$; *then, for* $\alpha, \beta \in \mathbb{N}$, $\alpha \neq 0$, *the difference*

$$\mathrm{length}_{W(k)}(M/F^{\alpha}M + p^{\beta}M) - \alpha\omega(\beta/\alpha)$$

is bounded.

We can replace k by $\bar{k}$, hence suppose k algebraically closed.

If M and M' are two lattices giving isomorphic F-spaces, there exists an exact sequence of W(k)[F]-modules

$$0 \longrightarrow M \longrightarrow M' \longrightarrow N \longrightarrow 0$$

where N has finite length. The snake-lemma, applied to the diagram

$$\begin{array}{ccccccccc} 0 & \longrightarrow & M\times M & \longrightarrow & M'\times M' & \longrightarrow & N\times N & \longrightarrow & 0 \\ & & \downarrow{\scriptstyle\varphi} & & \downarrow{\scriptstyle\varphi} & & \downarrow{\scriptstyle\varphi} & & \\ 0 & \longrightarrow & M & \longrightarrow & M' & \longrightarrow & N & \longrightarrow & 0 \end{array}$$

where $\varphi(x,y) = F^{\alpha}x + p^{\beta}y$, gives the inequality

$$\mathrm{length}\ M/\varphi(M^2) - \mathrm{length}\ M'/\varphi(M'^2) \leq 2\ \mathrm{length}\ N;$$

therefore, if the proposition is true for M(resp. M'), it is true for M'(resp. M).

c) It is therefore sufficient to prove the proposition for the F-lattices M_k^{λ}. In that case, M has a basis $e_1,\dots,e_r$ with $Fe_1 = e_2,\dots,Fe_{r-1} = e_r$, $Fe_r = p^s e_1$; if $\alpha = ar + b$, $0 \leq b \leq r-1$, then

$$F^{\alpha} e_1 = p^{as} e_{b+1}, \ldots, F^{\alpha} e_{r-b} = p^{(a+1)s} e_1, \ldots, F^{\alpha} e_r = p^{(a+1)s} e_b,$$

and $F^{\alpha} M + p^{\beta} M$ is generated by

$$p^{\inf(as,\beta)} e_i, \; i = b+1, \ldots, r \quad \text{and} \quad p^{\inf((a+1)s,\beta)} e_j, j = 1, \ldots, b.$$

The length of the quotient is

$$\ell = (r - b) \inf (as, \beta) + b \inf ((a+1)s, \beta).$$

If $\beta \leqslant as$, then $\ell = r\beta$;
if $as \leqslant \beta \leqslant (a+1)s$, then $\ell = (r - b)\, as + b\beta$;
if $(a+1)s \leqslant \beta$ then $\ell = (r - b)\, as + b(a+1)s = \alpha s$.

On the other hand $\omega(\beta/\alpha) = r \inf (\beta/\alpha, \lambda)$, hence $\alpha\omega(\beta/\alpha) = \alpha r \inf (\beta/\alpha, s/r) = \inf (\beta r, \alpha s)$, and the proposition follows easily.

The slopes, slope sequence,..., for a p-divisible group G over k (not necessarily algebraically closed, nor even perfect) are defined as the corresponding object for the F-space $E(G \otimes_k \bar{k})$.

The slopes of G are in the interval $[0,1]$. The above proposition gives:

<u>Corollary</u>. <u>If</u> ω <u>is the slope function of the p-divisible group</u> G, <u>then, for</u> $\alpha, \beta \in \mathbb{N}, \alpha \neq 0$

$$\mathrm{rk}(\mathrm{Ker}\, F_G^{\alpha} \cap \mathrm{Ker}\, p^{\beta} \mathrm{id}_G) = p^{\alpha\omega(\beta/\alpha) + A(\alpha,\beta)}$$

<u>where</u> $A(\alpha, \beta)$ <u>is bounded</u>.

In particular $\omega(\lambda) = 0$ for $\lambda \leqslant 0$,

$$\omega(\lambda) = \lim_{\substack{\alpha \longrightarrow \infty \\ \alpha, \lambda\alpha \in \mathbb{N}}} \frac{1}{\alpha} \log_p (\mathrm{rk}(\mathrm{Ker}\, F_G^{\alpha} \cap \mathrm{Ker}\, p^{\lambda\alpha} \mathrm{id}_G)), \text{ for } \lambda \geqslant 0,$$

$$\omega(\lambda) = \dim G \quad \text{for} \quad \lambda \geqslant \text{height } (G).$$

6. <u>The characteristic polynomial of an endomorphism.</u>

If M is an F-lattice (resp. E is an F-space) and φ an endomorphism of M (resp. E), then the determinant $\det(\varphi)$ of φ is in $\mathbb{Z}_p$ (resp. $\mathbb{Q}_p$): if $n = \dim M$ (resp. $n = \dim E$), then $\Lambda^n \varphi$ is the multiplication by $\det(\varphi)$ and commutes with F; this implies $\det(\varphi)^{(p)} = \det(\varphi)$, hence the assertion. More generally, the characteristic polynomial

$$\det(\varphi - T\, \mathrm{id})$$

of φ is in $\mathbb{Z}_p[T]$ (resp. $\mathbb{Q}_p[T]$).

If φ is an endomorphism of M, then it is well-known that

$$\text{length } (M/\varphi(M)) = v(\det(\varphi)).$$

(Note that $v(0) = \infty$).

This applies for instance to the case of the F-lattice of a p-divisible group G, and gives for any endomorphism φ of a p-divisible group G

$$\mathrm{rk}(\mathrm{Ker}\,\varphi) = p^{v(\det M(\varphi))},$$

(where, <u>by convention</u>, $p^{\infty} = 0$, and $\mathrm{rk}(H) = 0$ if H is not finite).

If k is a finite field with $q = p^a$ elements, then F^a is $W(k)$-linear, hence is an endomorphism of the F-lattice M (resp. of the F-space E).

Theorem (Manin). Let k be a finite field with $q = p^a$ elements, E an F-space, $\bar{\mathbb{Q}}_p$ the algebraic closure of $\mathbb{Q}_p$, $w: \bar{\mathbb{Q}}_p \longrightarrow \mathbb{Q}$ the valuation such that $v(q) = 1$ (i.e. $v(p) = 1/a$); let

$$P(T) = \det(F_E^a - T\,\mathrm{id}) = \pi(\tau_i - T)$$

$\tau_i \in \bar{\mathbb{Q}}_p$. Then the slopes of E are the $w(\tau_i)$ (counted with their multiplicities

By replacing E by E(-n), which replaces τ_i by $q^n \tau_i$, and the slopes (λ_i) by (λ_{i+n}), we can suppose that E comes from an F-lattice M. By the above proposition, the slope function ω of E is determined by $\omega(\lambda) = 0$ if $\lambda \leqslant 0$, and for $\lambda \geqslant 0$

$$(*) \qquad \omega(\lambda) = \lim_{\substack{\alpha \longrightarrow 0 \\ \alpha, \lambda\alpha \in \mathbb{N}}} \frac{1}{\alpha} \text{ length } M((F^{a\alpha} M + q^{\lambda\alpha} M).$$

Note that $B(k) \subset \mathbb{Q}_p$. We can find a basis e_i of $E \otimes_{B(k)} \bar{\mathbb{Q}}_p$ such that the matrix of F^a in this basis is triangular with diagonal entries τ_i; as remarked in the proof of the proposition of n° 5, the right hand side of (*) is also equal to the analogous expression, M being replaced by the lattice N in $E \otimes_{B(k)} \bar{\mathbb{Q}}_p$ generated by the e_i. But $F^{a\alpha} e_i = \tau_i^{\alpha} e_i$, and

$$\text{length } N/(F^{a\alpha} N + q^{\lambda\alpha} N) = \sum \inf(\alpha\, w(\tau_i), \lambda\alpha).$$

This gives $\omega(\lambda) = \sum \inf(w(\tau_i), \lambda)$, whence the theorem.

7. Specialization of p-divisible groups.

If S is a scheme over $\mathbb{F}_p$, a p-divisible group G over S is a system (G_n, i_n) of finite locally free commutative group-schemes over S, together with homomorphisms $i_n : G_n \longrightarrow G_{n+1}$ with the properties given in Ch. III.

For each $s \in S$, the fibres (G_n) give a p-divisible group G_s.

<u>Theorem</u> (Grothendieck). *Let $s' \in S$ be a specialisation of s, ω (resp ω') the slope-function of G_s (resp. $G_{s'}$). Then $\omega' \geqslant \omega$. Equivalently, the Newton-polygon of $G_{s'}$ is above the Newton-polygon of G_s.*

Each $\operatorname{Ker} F_G^{\alpha}$ and each $\operatorname{Ker} p^{\beta} \mathrm{id}_G$ is a finite locally free commutative group scheme; moreover

$$\operatorname{Ker} F^{\alpha}, \operatorname{Ker} p^{\beta} \subset \operatorname{Ker} p^{\sup(\alpha,\beta)}.$$

By the following lemma, it follows that

$$\operatorname{rk}(\operatorname{Ker} F_G^{\alpha} \cap \operatorname{Ker} p^{\beta} \mathrm{id}_G)_{s'} \geqslant \operatorname{rk}(\operatorname{Ker} F_G^{\alpha} \cap \operatorname{Ker} p^{\beta} \mathrm{id}_G)_s.$$

This gives immediately $\omega_{s'}(\lambda) \geqslant \omega_s(\lambda)$.

<u>Lemma</u>. *Let S be a scheme, Z a finite locally free S-scheme, X and Y two finite locally free closed subschemes of Z. If $s' \in S$ is a specialisation of $s \in S$, then*

$$\operatorname{rk}(X \cap Y)_{s'} \geqslant \operatorname{rk}(X \cap Y)_s.$$

<u>Proof</u>. Take $S = \operatorname{Spec} R$ affine, $Z = \operatorname{Spec} A$, $X = \operatorname{Spec} A/I$, $Y = \operatorname{Spec} A/J$; then $X \cap Y = \operatorname{Spec} A(I + J)$. But A/I and J are locally free R-modules and $A/(I + J)$ is the cokernel of the R-linear map $\varphi : J \longrightarrow A/I$. Remark now that the rank of φ_s does not increase by specialization.

<u>Remark</u>. If G is of <u>height</u> r, then $\omega_s(r)$ is the dimension of G_s. Hence $\omega_{s'}(Z) = \omega_s(Z)$; equivalently, the extremities of the Newton polygon are invariant under specialization.

8. <u>Some particular cases</u>.

Let G be a p-divisible group (k perfect). The slope sequence of G:

$$\lambda_1 \leq \lambda_2 \leq \cdots \leq \lambda_h \quad \text{with} \quad 0 \leq \lambda_1, \lambda_k \leq 1,$$

determines $G \otimes_k \bar{k}$ upto isogeny. We know that G splits as a product $G_e \times G_c$, where G_e is etale and G_c connected. But G^{λ} is etale (resp. connected) if and only if $\lambda = 0$ (resp $\lambda > 0$). Hence the slopes of G_e (resp. G_c) are the λ_i which are $= 0$ (resp. > 0).

The Serre dual G' of G has the slope sequence

$$1 - \lambda_h \leq 1 - \lambda_{h-1} \leq \cdots \leq 1 - \lambda_1 .$$

Applying the preceeding decomposition also to G', we find:

<u>Proposition</u>. <u>The p-divisible group G can be uniquely written as a product</u>

$$G = G_e \times \bar{G} \times G_m,$$

<u>where the slopes of</u> G_e (resp. $\bar{G}$, resp. G_m) <u>are the slopes of</u> G <u>which are</u> $= 0$ (resp $\neq 0, 1$, resp $= 1$).

In particular, if $k = \bar{k}$, then $G_e = (\mathbb{Q}_p/\mathbb{Z}_p)_k^{m_0}$, $G_m = (\mu_k(p))^{m_1}$.

<u>Proposition</u>. <u>If</u> G <u>is isogenous to</u> $G_k^{1/r}$ (resp. $G_k^{(r-1)/r}$), <u>then</u> G <u>is isomorphic to it</u>.

Equivalently: if $\lambda = 1/r$, or $(r-1)/r$, any F-lattice M in E_k^{λ} is isomorphic to M_k^{λ}. By duality, it is enough to prove the statement for $\lambda = 1/r$. Then E_k^{λ} has a basis $e_1, \ldots, e_r$, with $Fe_1 = e_2$, $Fe_2 = e_3, \ldots, Fe_{r-1} = e_r$, $Fe_r = pe_1$. For each i, let $m_i = \inf\{m \mid p^m e_i \in M\}$. Then

$$m_1 \geq m_2 \geq \cdots \geq m_n \geq m_1 - 1;$$

replacing if necessary the basis (e_i) by a basis $(F^{\alpha} p^{\beta} e_i)$, we can suppose that $m_1 = m_2 = \cdots = m_n = 0$, i.e. $e_i \in M$ and $p^{-1} e_i \notin M$, for all i. This implies $M \supset M_k^{\lambda}$. Let $m \in M$, $m \notin M_k^{\lambda}$; write

$$m = \sum a_i e_i, \quad a_i \in B(k).$$

There exists α with $F^{\alpha} m \notin M_k^{\lambda}$, $F^{\alpha+1} m \in M_k^{\lambda}$; replacing m by $F^{\alpha} m$, we can suppose $m \notin M_k^{\lambda}$, $m \in M$, $Fm \in M_k^{\lambda}$; but

$$Fm = pa_n e_1 + a_1 e_2 + \cdots + a_{n-1} e_n,$$

hence $a_1, \ldots, a_{n-1} \in W(k)$, $a_n \notin W(k)$, $pa_n \in W(k)$. This implies

$$a_n e_n = Fm - a_1 e_2 \cdots - a_{n-1} e_n \in M,$$

and a contradiction.

<u>Example</u>. If $k = \bar{k}$, then any p-divisible group G of height $0,1,2,3$ is isomorphic to one of the following:

height 0 : 0

height 1 : G_0, G_1

height 2 : G_0^2, G_1^2, $G_0 \times G_1$, $G_{1/2}$.

height 3 : G_0^3, $G_0^2 \times G_1$, $G_0 \times G_1^2$, G_1^3, $G_{1/3}$, $G_{2/3}$.

For height 4, it is isomorphic to G_0^4, $G_0^3 \times G_1$, $G_0^2 \times G_1^2$, $G_0 \times G_1^3$, G_1^4, $G_0^2 \times G_{1/2}$, $G_0 \times G_1 \times G_{1/2}$, $G_1^2 \times G_{1/2}$, $G_{1/4}$, $G_{3/4}$, or isogenous to $(G_{1/2})^2$.

CHAPTER V

p-ADIC COHOMOLOGY OF ABELIAN VARIETIES

k is a field, $p = \text{charac}(k)$.

1. Abelian varieties, known facts.

The following facts are known, see Lang's or Mumford's Abelian Varieties.

a) If A is an abelian variety, of dimension g, over k, and $\varphi_1, \ldots, \varphi_r$ are endomorphisms of A, then

$$\text{rk Ker}(n_1 \varphi_1 + \ldots + n_r \varphi_r)$$

is a polynomial in $n_1, \ldots, n_r$ with rational coefficients, homogeneous of degree $2g$ (by convention, the rank of a non-finite group is 0).

For instance, $\text{rk Ker}(n\,\text{id}_G) = n^{2g}\,\text{rk Ker}(\text{id}_G) = n^{2g}$.

The characteristic polynomial P of the endomorphism φ is defined by

$$P(n) = \text{rk Ker}(\varphi - n\,\text{id}_G) = (-1)^{2g}\, n^{2g} + \cdots \text{rk}(\text{Ker}\,\varphi).$$

b) There exists an abelian variety A', the dual of A, with the following properties:

1) for any n, $\text{Ker}(n\,\text{id}_A)$ and $\text{Ker}(n\,\text{id}_{A'})$ are Cartier dual of each other, this duality being compatible with the inclusion and projection

$$\text{Ker}(n\,\text{id}_A) \longrightarrow \text{Ker}(nm\,\text{id}_A) \xrightarrow{\;n\;} \text{Ker}(m\,\text{id}_A),$$

$$\text{Ker}(n\,\text{id}_{A'}) \xleftarrow{\;m\;} \text{Ker}(nm\,\text{id}_A) \longleftarrow \text{Ker}(m\,\text{id}_{A'}).$$

2) There exists an isogeny (epimorphism with finite kernel) of A to A'.

2. Points of finite order and endomorphisms.

Let A be an abelian variety over k, and ℓ a prime number. For any $n \in \mathbb{N}$, $\mathrm{Ker}(\ell^n \, \mathrm{id}_A)$ is a finite group of rank ℓ^{2ng}. We define

$$A(\ell) = \bigcup_n \mathrm{Ker}(\ell^n \, \mathrm{id}_A).$$

$$A(\ell) \otimes_k \bar{k} \simeq (\mathbb{Q}_\ell(\mathbb{Z}_\ell)^{2g}.$$

If $\ell \neq p$, then $A(\ell)$ is an etale formal-group, and we define

$$H^1(A, \ell) = \mathrm{Hom}_{\mathbb{Z}_\ell}(A(\ell) \otimes_k \bar{k}, \mathbb{Q}_\ell/\mathbb{Z}_\ell);$$

it is a free module of rank 2g over $\mathbb{Z}_\ell$ (and also a Galois module).

If $\ell = p$, then $A(p)$ is a p-divisible group, of height 2g. We define

$$H^1(A,p) = M(A(p)) = \text{Dieudonné module of } A(p);$$

it is an F-lattice over k, and in particular a free module of rank 2g over W(k).

Evidently $A \longmapsto H^1(A, \ell)$, ℓ any prime, is a functor. In particular, any endomorphism φ of the abelian variety A gives rise to an endomorphism $H^1(\varphi, \ell)$ of $H^1(A, \ell)$. We denote by v_ℓ the canonical valuation on $\mathbb{Z}_\ell$ (resp. W(k) if $\ell = p$).

Lemma. If φ is an endomorphism of A, then for any prime ℓ, ($\ell \neq p$ or $\ell = p$), the highest power of ℓ which divides $\mathrm{rk}(\mathrm{Ker}\, \varphi)$ is $\ell^{v_\ell(\det H^1(\varphi, \ell))}$.

Equivalently

$$v_\ell(\mathrm{rk}(\mathrm{Ker}\, \varphi)) = v_\ell(\det(H^1(\varphi, \ell)).$$

We can suppose k is algebraically closed. As we have seen, Ker φ is the product of its components of ℓ-torsion:

$$\operatorname{Ker} \varphi = \prod (\operatorname{Ker} \varphi \cap A(\ell))$$

and $\operatorname{rk}(\operatorname{Ker} \varphi \cap A(\ell))$ is a power of ℓ, hence

$$\operatorname{rk}(\operatorname{Ker} \varphi \cap A(\ell)) = \ell^{v_\ell (\operatorname{rk} \operatorname{Ker} \varphi)}.$$

For each ℓ, φ induces an endomorphism of $H^1(A, \ell)$ and we have an exact sequence

$$H^1(A, \ell) \xrightarrow{H^1(\varphi, \ell)} H^1(A, \ell) \longrightarrow N \longrightarrow 0,$$

where N is of length $v_\ell(\det H^1(\varphi, \ell))$.

If $\ell \neq p$, N is the Pontrjagin dual of $\operatorname{Ker} \varphi \cap A(\ell)$, hence the relation. If $\ell = p$, N is the Dieudonné module of $\operatorname{Ker} \varphi \cap A(p)$, and $\operatorname{rk}(\operatorname{Ker} \varphi \cap A(p)) = p^{\text{length } (N)}$ as we have seen.

Theorem. *If φ is an endomorphism of A, then, for any ℓ, ($\ell \neq p$, or $\ell = p$), we have*

$$\operatorname{rk}(\operatorname{Ker} \varphi) = \det H^1(\varphi, \ell).$$

This follows from the preceding lemma, by the method of Mumford, p. 181.

Corollary. *If φ is an endomorphism of A, then the characteristic polynomial of φ is also the characteristic polynomial of $H^1(\varphi, \ell)$ for all ℓ. It has integral coefficients.*

Because a rational number is integral if it is a ℓ-adic integer for all ℓ.

3. <u>Structure of the p-divisible group</u> $A(p)$.

We remark first that $A'(p)$ (A' the dual abelian variety to A) is canonically isomorphic to the Serre dual of $A(p)$. Because A' and A are isogenous, this implies that $A(p)$ is isogenous to its Serre dual. Equivalently, if the <u>slope sequence</u> of $A(p)$ is

$$\lambda_1 \leq \lambda_2 \leq \cdots \leq \lambda_{2g},$$

then $\lambda_i + \lambda_{2g-i} = 1$.

<u>Remark</u>. If $\lambda_i = \frac{s}{r}$, then $\lambda_{2g-i} = \frac{r-s}{r}$, and $s + (r-s) = r$. From these follows the well-known fact that the <u>dimension</u> of $A(p)$ is g, i.e. $\mathrm{rk}(\mathrm{Ker}\, F_A^i) = p^{ig}$.

For instance, if $g = 1$, then $A(p) \otimes_k \bar{k}$ is isogenous (hence isomorphic) to either $G_0 \times G_1$ or $G_{1/2}$. More generally:

<u>Proposition. Let</u> A <u>be an abelian variety of</u> dim g, <u>over the algebraically closed field</u> k. <u>Then</u> $A(k)$ <u>contains at most</u> p^g <u>points of order</u> p. <u>Moreover, the following conditions are equivalent.</u>

1) $A(k)$ <u>contains</u> p^g <u>points of order</u> p.

2) $A(p)$ <u>is isomorphic to</u> $G_0^g \times G_1^g$.

3) $\mathrm{Ker}(p\, \mathrm{id}_G : A \longrightarrow A)$ <u>is isomorphic to</u> $(\mathbb{Z}/p\mathbb{Z})_k^g \times (p^{\mu}k)^g$.

We have $A(p) = \hat{A}^0 \times (\mathbb{Q}_p/\mathbb{Z}_p)_k^r$; the slopes of $\hat{A}^0$ are the $\lambda_i > 0$, the slopes of $(\mathbb{Q}_p/\mathbb{Z}_p)^r$ are the $\lambda_i = 0$. Hence r is the multiplicity of the slope 0, hence also the multiplicity of the slope 1. This implies $r \leq g$, and the equivalence $r = g \Longleftrightarrow$ the slopes of $A(p)$ are g times 0 and g times 1. The proposition follows easily.

Such an abelian variety is called *ordinary*.

The theorems of § 2 and Chapter IV, § 6 give:

Theorem (Manin). *Let* k *be a finite field with* $q = p^a$ *elements,* A *an abelian variety over* k,

$$P(T) = \sum_{i=1}^{2g} (\tau_i - T) = T^{2g} + \cdots + q^n$$

$\tau_i \in \bar{\mathbb{Q}}_p$, *the characteristic polynomial of the Frobenius endomorphism* F_A^a *of* A. *Then the slopes of* $A(p)$ *are* $w(\tau_i)$ *where* w *is the valuation* $\bar{\mathbb{Q}}_p \to \mathbb{Q}$ *such that* $w(q) = 1$.

Example. If $g = 1$, i.e. A is an elliptic curve, then

$$P(T) = T^2 - \mathrm{Tr}(F^a) + q,$$

and we find the (easy) statements:

$$\mathrm{Tr}(F^a) \equiv 0 \pmod p \iff A(p) = G_{1/2}$$

$$\mathrm{Tr}(F^a) \not\equiv 0 \pmod p \iff A(p) = G_0 \times G_1 \quad \text{i.e. } A \text{ is ordinary.}$$